? JUN 2009

'0/13

The **Astronomy**
H A N D B O O K

The Astronomy

HANDBOOK

GUIDE TO THE NIGHT SKY

Clare Gibson

Published by SILVERDALE BOOKS
An imprint of Bookmart Ltd
Registered number 2372865
Trading as Bookmart Ltd
Blaby Road
Wigston
Leicester LE18 4SE

© 2005 D&S Books Ltd

D&S Books Ltd
Kerswell,
Parkham Ash, Bideford
Devon, England
EX39 5PR

e-mail us at:- enquiries@d-sbooks.co.uk

This edition printed 2005

ISBN 1-84509-001-2

DS0082. Astronomy Handbook

All rights reserved. This book is protected by copyright. No part of it may be
reproduced, stored in a retrieval system, or transmitted in any form or by any means,
without the prior permission in writing of the Publisher, nor be circulated in any form
of binding or cover other than that in which it is published and without a similar
condition including this condition being imposed on the subsequent Publisher

Creative Director: Sarah King
Editor: Sally MacEachern
Project editor: Nicola Barber
Designer: Big Metal Fish

Printed in Singapore

1 3 5 7 9 10 8 6 4 2

NORTHAMPTONSHIRE LIBRARIES	
802679107	
Bertrams	25.07.07
520	£4.99
034395	WW

CONTENTS

Solar eclipses ha e inspired awe for millennia.

Introduction

> *Before the days of Kepler the heavens declared the glory of the Lord.*
>
> George Santayana,
> *The Sense of Beauty* (1896)

THE HISTORY OF ASTRONOMY

The science of astronomy dates back for millennia, to the days when the Sun, Moon, planets and stars, along with the celestial sphere that they inhabit, were believed to be divine entities and immortal beings. Although we now know differently, this unsophisticated viewpoint is perfectly understandable when you consider the seemingly miraculous way in which the Sun's energising light prompts seeds to germinate and then rapidly to transform themselves into crops that can be harvested and eaten, thereby sustaining human life; how the Moon's influence regulates the tides of the oceans and seas and women's menstrual cycles; and how magical the twinkling stars look against the dark, velvety, night sky. In ancient Greece, to take just one example, the Sun was thought to be the manifestation of the god Apollo, while the Moon, with its distinct phases, was associated with the triple goddess, or with three goddesses of very different aspects, namely the maiden Artemis, the motherly Selene and implacable Hecate, the goddess of the underworld. With the blazing Sun dominating the sky by day, and the mysterious Moon coming into its own at night, the former came to be equated with vitality and the life force, and the latter with fertility and the gentle nurturing and perpetuation of that life force, yet also with death. And can you imagine

To ancient cultures, the Moon's influence on the tides would have seemed miraculous.

Historians believe that such ancient monuments as the Pyramids (above) and Stonehenge (above right) served as solar calendars, and that they were also used in sun-worshipping rites.

the terror that a solar eclipse – when the sky darkens, birds fall silent and the air becomes increasingly chilly as the Moon appears to devour the Sun – must have aroused in our early ancestors?

In an age when life was a constant struggle for survival, and when a drought or a flood could decimate a population, it follows that humans believed themselves to be absolutely at the mercy of the Sun's will, in particular. This is why some cultures, such as that of the Aztecs, attempted to persuade their solar god (the Aztecs' was Huitzilopochtli) to look kindly on them by practising human sacrifice in his honour. The Aztecs regularly performed this bloody ritual in Tenochtitlan, at the summit of Templo Mayor, a shrine erected atop a manmade pyramid similar to those constructed by the ancient Egyptians. Although, like that of Stonehenge in England, their exact significance remains a mystery, historians believe that such monumental structures had profound cosmological symbolism and furthermore acted as solar markers, or as calendars in stone, and consequently as backdrops for Sun-worshipping rites.

Many early peoples – including the Mayans and Aztecs, the Egyptians and Babylonians, the Chinese and Indians, to name but a few dedicated star-gazers – placed great importance on monitoring the night sky and recording any phenomena that they observed there. This was partly because discerning patterns, such as the changing appearance of the Moon during the course of the month, and that of the Sun throughout the year, as well as charting the positions of the stars in relation to one another, served several useful purposes – to determine the best times for planting, or for orientation while travelling, for instance. And partly because any aberrations from the norm were regarded as portents of important future events, be they wonderful or terrible. The New Testament's Gospel according to St Matthew, for example, relates that three wise men from the East – most likely Chaldean astronomers – appeared before King Herod, asking, 'Where is he that is born King of the Jews? For we have seen his star in the east, and are come to worship him' (2:2). Matthew then tells us that when the magi later set off for Bethlehem to salute the newborn Jesus Christ, '. . . the star, which they saw in the east, went before them, till it came and stood over where the young child was' (2:9). This snippet of biblical testimony may, in fact, be one of the earliest allusions to Halley's Comet, a periodic comet whose appearance was first recorded by Babylonian astronomers in 164 BC.

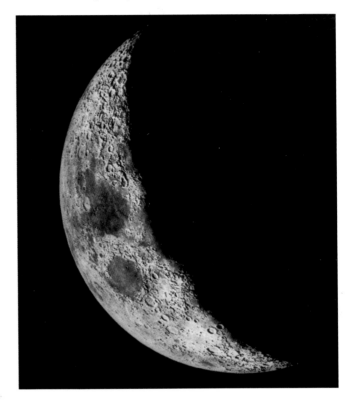

Early peoples placed great importance on recognising patterns of change in the night sky, such as the waxing and waning of the Moon. Monitoring such changes was important for crop-growing and marking the seasons.

THE OLDEST SCIENCE

The twin roots of the name of the world's oldest science – astronomy – are the Greek words *astron*, which means 'star', and *nomia*, 'law'. Although it was the Greeks who, as the designation of the years changed from BC to AD, were responsible for formulating and then disseminating the laws that they believed governed the behaviour of the celestial bodies, they were building on groundwork laid primarily by Babylonian observers of the heavens (who had compiled the first almanacs, or collections of astronomical statistics, by around 750 BC), and by Egyptians, too. Even so, significant advances in humankind's understanding of matters astronomical were made in ancient Greece, and such Greek names as Aristarchus of Samos (*c.* 320–250 BC), who calculated the distance between the Sun and Moon and also suggested that the Earth orbited the Sun, and Hipparchus (*c.* 190–120 BC), who, in around 130 BC, measured the positions of 850 stars and catalogued them, classifying them by brightness on a scale of one to six, and who also identified the precession of the Earth's axis, certainly merit places in astronomy's hall of fame.

For one-and-a-half millennia, the shining star of Greek astronomy was, however, Ptolemy (*c.* AD 90–169). Actually an Alexandria-based Egyptian rather than a Greek, Ptolemy's fame was mainly due to his hugely influential book

*Portrait plaque depicting Ptolemy,
Egyptian, 305-50 BC (gold).*

The Almagest, although he was also
celebrated for the accuracy of his maps of
the known world. Comprising as it did a
mixture of his own and many of his
predecessors' theories, findings and
conclusions, *The Almagest* served as the
definitive astronomy handbook in the West
until the 17th century, in the process
perpetuating a critical fallacy, however,
namely that the universe is geocentric. For
according to *The Almagest*, the Sun and
planets – which Ptolemy specified as being
the Moon (again, as we now know,
wrongly), Mercury, Venus, Mars, Jupiter and
Saturn – orbit the Earth within concentric
spheres, the eighth, and outermost, sphere
being home to the fixed stars.

*Aristarchus of Samos, the ancient Greek
astronomer, was one of the first to
suggest that the Earth orbited the Sun.*

THE ISLAMIC CONTRIBUTION

The Arabs followed the ancient Greeks in picking up the baton of astronomical endeavour. Apart from keeping the science of astronomy alive while Europe underwent its Dark Ages, one of their most important contributions to the furtherance of humankind's knowledge of the heavens was the development of the astrolabe and other measuring devices. The astronomers of ancient Greece had been aided in determining the positions of stars by the armillary sphere, an instrument that consisted of a series of moveable concentric rings, each signifying a planet and together representing the celestial sphere, and a sighting device. The altitude-measuring quadrant – which comprised a 90° arc with graduations used in conjunction with a sighting device attached to a moveable arm – was also used in ancient Greece, and this the Arabs successfully improved and enlarged, thereby increasing its accuracy. The astrolabe measured altitude as well, and made astronomers' lives considerably easier by performing calculations for them, thanks to an integral, pierced-metal star map and sighting device that could be rotated to reflect different latitudes. The successful honing of these instruments' efficiency by around AD 900 enabled the sky-watching Arabs to create the most detailed and precise stellar representations and almanacs yet produced.

Nicolaus Copernicus (1473–1543), the Polish astronomer who abandoned the Ptolemaic System (according to which the Earth was the centre of the universe) and instead worked out a heliocentric model (in which the Sun is the centre of the universe).

THE COPERNICAN REVOLUTION

History does not relate whether a comet streaked through the sky above his Polish birthplace when Nicolaus Copernicus was born in 1473, but he is certainly remembered as a trailblazer. Indeed, such is his significance in the history of astronomy that, like Ptolemy, his name describes an astronomical model, the difference being that while the Ptolemaic system is geocentric and erroneous, the Copernican system is heliocentric and fundamentally correct. (There was a flaw in Copernicus' heliocentric theory, however, namely his assumption that the Earth and planets follow a perfectly circular course in their orbits around the Sun.) Not that it was immediately accepted as such, however, when, 30 years after its conception, Copernicus at last made public his theory that the Earth was just one of a number of planets that revolved around the Sun, and that it was therefore not the centre of the universe. As a lay canon at Frauenberg (now Frombork) Cathedral, Copernicus must have

A page from a Dutch treatise, published in Latin in 1617, which discusses Copernicus' theory that the Earth circles the Sun, and not the other way around.

THE CREATION OF THE SUN AND MOON, FROM THE OLD TESTAMENT

14 And God said, Let there be lights in the firmament of the heaven to divide the day from the night; and let them be for signs, and for seasons, and for days, and years:

15 And let them be for lights in the firmament of the heaven to give light upon the earth: and it was so.

16 And God made two great lights; the greater light to rule the day, and the lesser light to rule the night: he made the stars also.

17 And God set them in the firmament of the heaven to give light upon the earth,

18 And to rule over the day and over the night, and to divide the light from the darkness: and God saw that it was good.

Genesis.

anticipated accusations of blasphemy from the all-powerful Roman Catholic Church, which is no doubt why he waited until he was dying before publishing *De Revolutionibus Orbium Coelestrium* (*Of the Revolutions of the Stellar Orbs*) in 1543.

Tycho Brahe (1546–1601), a Dane and the leading astronomer of his age, reconciled the conflicting Copernican and Ptolemaic systems in an ingenious fashion by proposing that the planets, apart from the Earth, revolved around the Sun, and that the Sun and Moon – the Sun all the while surrounded by its circling satellites – in turn revolved around the Earth. Although this hybrid theory was incorrect, astronomy has cause to be grateful to Brahe, who was employed first by King Frederick II of Denmark, from whom, in 1576, he received the island of Hven, in Öre Sound, on which to build an observatory, and then by King Rudolf II of Bohemia, which is why he spent his last few years in Prague. During the 20 years that he systematically observed the sky from Hven, and with the help of excellent instruments, Brahe meticulously calculated the positions of the Sun and planets in relation to the stars. He also recorded any unusual events that he saw occurring overhead, and is especially remembered for

A map of Tycho Brahe's system of planetary orbits around the Earth, which he created between 1660 and 1661.

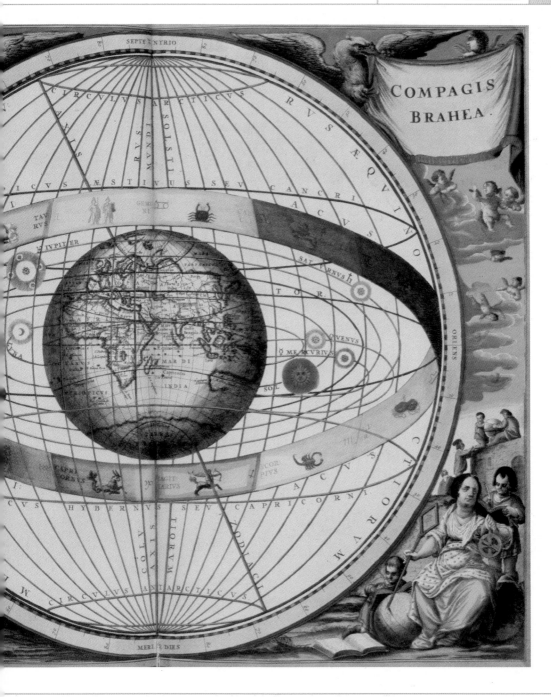

his observations, between 1572 and 1574 and in 1577 respectively, firstly of the appearance of a new star (which we now know to be a supernova) in the constellation of Cassiopeia, and secondly, of the elongated orbit of 'Tycho's Comet' as it travelled past the planets beyond the Moon. Brahe had provided crucial evidence that the stars in the sky can change, and that comets are not within Earth's atmosphere, contrary to previous belief.

Brahe was also notable for employing the German astronomer Johannes Kepler (1571–1630) to act as his assistant shortly before his death, and especially for providing his heir in astronomy with the raw data – in the form of his observational records – that enabled the formulation of Kepler's laws of planetary motion, of which there are three. First, planets orbit the Sun in an elliptical fashion, with the Sun being positioned at one of the two foci. Second, an imaginary line (a radius vector) joining a planet

Johannes Kepler (1571–1630), the German astronomer, is depicted discussing his discoveries of planetary motion with his patron, Rudolph II of Bohemia.

and the Sun during the planet's orbit sweeps out over equal areas in equal times. And, third, the squares of two planets' periods are proportional to the cubes of their orbital major axes, or their distance from the Sun. With the publication of the first two laws in his work *Astromia Nova* (*New Astronomy*) in 1609, and of the third in *Harmonices Mundi* (*Harmonies of the World*) in 1619, Kepler both confirmed and corrected Copernicus' heliocentric theory, thereby completing the Copernican revolution.

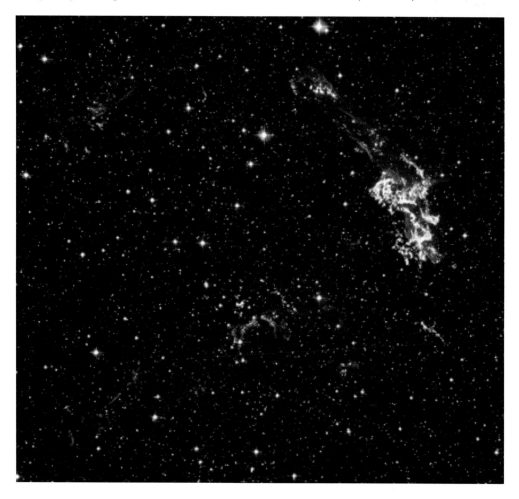

This image presents a view of Kepler's supernova remnant taken in X-rays, visible light and infrared radiation by the Hubble Space Telescope.

A portrait of the Italian astronomer Galileo Galilei (1564–1642).

THE NEW 'WONDER INSTRUMENT'

'Oh thou tube of knowledge, more precious than any sceptre', exclaimed Kepler in his book *Dioptrice* (*Dioptrics*), which was published in 1611. Kepler was referring to the telescope, the 'wonder instrument', as he called it, that the Dutch spectacle-maker Hans Lippershey (*c.* 1570–*c.* 1619) is thought to have invented in 1608. Word of Lippershey's invention spread like wildfire, inspiring others to emulate his example and build their own telescopes. One of them was Galileo Galilei (1564–1642), an Italian mathematician, physicist and astronomer, who, in 1609, constructed a refracting telescope whose lenses gave it a magnification of around 8x, which he later improved to 30x.

What Galileo saw when he trained his telescope on the heavens, not least the innumerable individual stars within the Milky Way, amazed and excited him. The discoveries that he subsequently made in turn rocked his readership when he published news of them in his work *Siderus Nuncius* (*Starry Messenger*) in 1610.

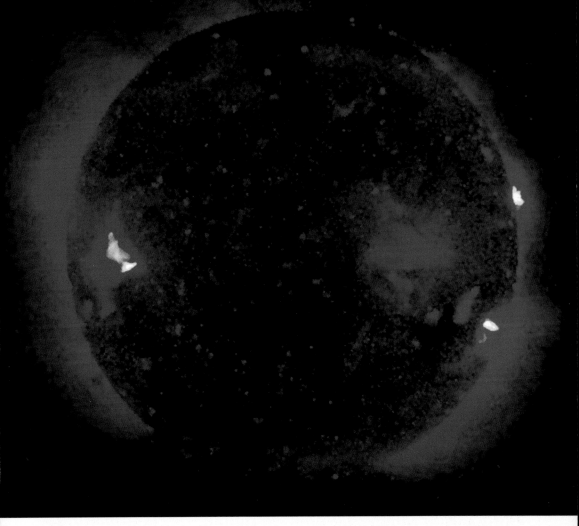

Among Galileo's discoveries were craters on the Moon (far left), four moons ('stars') orbiting Jupiter (left) and sunspots (above).

Galileo reported seeing mountains and craters on the Moon, as well as four 'stars' orbiting the planet Jupiter, which he eventually concluded must be satellites or moons. Further discoveries made in 1610 included sunspots and the lunar-like phases exhibited by the planet Venus. Galileo was convinced that he had now seen proof of the Copernican view of the universe with his own eyes, but, mindful of the Church's implacable opposition to the heliocentric theory, delayed publishing his cosmological work *Dialogue on the*

Huygen's discoveries regarding
Saturn included a ring around it
(above), and its largest moon,
Titan (below).

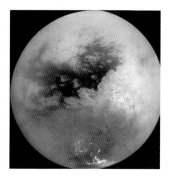

Two Great World Systems until 1632. Horrified by its
contents, the Church banned the book, subjected its author to
house arrest and eventually 'persuaded' Galileo to recant his
allegedly heretical, Copernican views. Only in 1979 did the
Vatican officially concede that the Earth revolved around the
Sun, and not *vice versa*.

Whatever else it had the power to inhibit in Catholic
countries such as Italy, the Church could not quash scientists'
desire to learn more about the mysteries of the sky above, nor
could it ban the use of telescopes, which astronomers
continued to refine and improve. Indeed, maybe it was the
eyepiece micrometer that he devised for his telescope that, in
1655, enabled the Dutch astronomer Christiaan Huygens
(1629–95) to ascertain that Saturn had a satellite (Titan).
Huygens later also discerned a ring circling the planet.
Building on the Dutchman's work, in 1675 the Italian
Giovanni Cassini (1625–1712) went on to identify a gap in

The gap in Saturn's rings, or more precisely between rings 'A' and 'B', is now named after its discoverer, Giovanni Cassini.

Saturn's rings (now named the Cassini Division in his honour) and also four more of Saturn's satellites.

Huygens and Cassini both relied on refracting telescopes to study Saturn more closely, but these instruments had an inherent, lens-created problem, namely spherical aberration, which meant that the image that they produced by refracting, or bending, light rays was not that sharp. Longer focal lengths could compensate for this aberration to some extent, but necessitated a hugely long, heavy and unwieldy tube. In 1663, James Gregory (1638–75), a Scottish mathematician, invented the first reflecting telescope, in which two concave mirrors reflected, or bounced, light back and forth within the tube, and ultimately to the eyepiece at the centre of one of the mirrors. The substitution of mirrors for lenses produced a sharper image.

It was the English scientist Isaac Newton (1643–1727) whose name is more popularly associated with the reflector, however, because it was his design that was generally adopted by the astronomical community. While studying the nature of light during the 1660s, Newton found that when white light passes through a lens or prism, it is separated into a spectrum of colours, resulting in an image with a rainbow fringe, or

A portrait of Sir Isaac Newton.

Newton's work on the refraction of light led to his construction of one of the first reflecting telescopes.

Newton's discoveries are still being utilised in modern telescopes today.

chromatic aberration. (*Opticks*, published in 1704, contains his optical findings and conclusions.) Suspecting that using mirrors in place of lenses would do away with this unwanted side effect, and would also allow a larger aperture, in 1668 Newton put his theory to the test and built his first refracting telescope, within which a large, concave mirror reflected the light to an angled plane mirror, which in turn reflected it to a hole in the side of the tube that functioned as the eyepiece. Success! Following the demonstration of his reflecting telescope to members of the Royal Society in London in 1672, astronomers enthusiastically adopted the Newtonian telescope. Telescopes of this type still aid astronomers to this day, as do refracting telescopes (see pages 212–16).

The Newtonian telescope is not the only reason why astronomers have cause to be thankful to Isaac Newton. Indeed, he is most famous for his theory of gravitation (gravity) and three laws of motion (of a body). Briefly, Newton's law of gravitation states that two particles attract one another with forces directly proportional to the product of their masses divided by the square of the distance between them. And Newton's first law of motion holds that a body remains at rest, or in uniform motion in a straight line, unless acted on by a force; his second law of motion states that a body's rate of change of momentum is in proportion to the force causing it; and, finally, his third law of motion maintains that when a force acts on a body, an equal and opposite force acts on another body at the same time. These theories are of crucial significance in astronomy because they explain, for example, why planets rotate around the Sun and their satellites in turn orbit the planets.

It was his friend Edmond Halley (1656–1742), a fellow English mathematician and member of the Royal Society, who persuaded Newton to share his theories with the wider world, which he eventually did in his work *Philosophiae Naturalis Principia Mathematica* (*Mathematical Principles of Natural Philosophy*), published in 1687. Halley made some notable contributions to the growing canon of astronomical knowledge himself, such as proving that some comets are periodic. He correctly calculated that one particular comet returns to the Sun every 76 years. When it returned, as he had foretold, in 1758, it was named after him.

Edmond Halley successfully predicted the periodic return of the comet that was later named after him.

EXPLORATION, DISCOVERY AND CONSOLIDATION

If the Copernican system gave Enlightenment-era astronomers their rules of engagement, increasingly sophisticated telescopes enabled them to see the stars, planets and heavenly phenomena in better detail. Discernment and discovery go hand in hand, and as one revelation followed another over the next few centuries, Earthlings found themselves being forced to reassess their existing view of first the solar system, and then the universe. One night in 1781, for instance, the German-born organist William Herschel (1738–1822) set up his telescope outside his house in Bath, England, in order to indulge in his hobby of star-gazing, and went to bed having discovered the planet Uranus (which he initially believed to be a comet). And between them, the Frenchman Urbain Le Verrier (1811–77) and the German Johann Galle (1812–1910) found Neptune. Le Verrier's achievement of 1845 was to predict the position

Below left: William Herschel discovered Uranus, although he originally believed it to be a comet.

Below: Neptune was discovered by French and German astronomers Urbain Le Verrier and Johann Galle.

Below right: Percival Lowell and Clyde Tombaugh are credited with the discovery of Pluto, Tombaugh having followed Lowell's instinct to pinpoint the elusive planet.

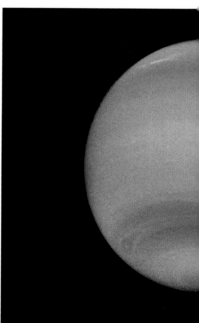

of a planetary body that appeared to be influencing Uranus'
orbit, and Galle's triumph was in proving Le Verrier correct
in 1846. The discovery of Pluto has similarly been credited
to two people: Percival Lowell (1855–1916), the American
astronomer who, in 1905, predicted – inaccurately, as it
happens – the position of Pluto, having concluded that an
unknown planet, which he called 'Planet X', was exerting an
influence on Uranus and Neptune's orbits; and the American
Clyde Tombaugh (1906–97), who, in pursuing Lowell's
hunch, pinpointed Pluto's position in 1930.

At the same time as advancing human knowledge of the
sky above, astronomers were busily consolidating it. For
example, not only did Charles Messier (1730–1817), the
French 'Comet Ferret', as he was dubbed, discover 15
comets, but he also compiled a list of 103 fuzzy-looking
objects that he had spotted overhead to prevent him from
confusing them with his beloved comets. Published in 1784,

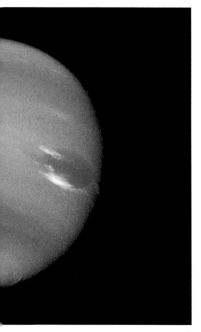

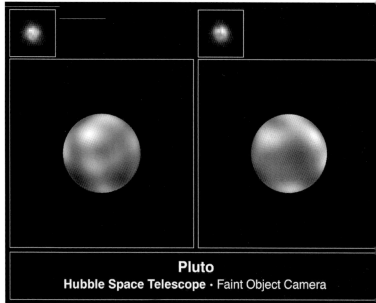

Pluto
Hubble Space Telescope · Faint Object Camera

PRC96-09a · ST ScI OPO · March 7, 1996 · A. Stern (SwRI), M. Buie (Lowell Obs.), NASA, ESA

the Messier Catalogue survives today in the form of the 'M' (for 'Messier') numbers that are the alternative names of certain nebulae, star clusters and galaxies, such as M1 (or the Crab Nebula), M67 (a star cluster) and M31 (which is also known as the Andromeda Galaxy).

The many other individuals who have contributed vital pieces to the vast, and still uncompleted, jigsaw puzzle that makes up our picture of the universe are too numerous to mention in this brief history of astronomy, but you will come across some of their names in connection with a stellar achievement in the pages that follow. These human stars of astronomy include dedicated female astronomers such as the Americans Annie Jump Cannon (1863–1941), who categorised stars by measuring absorption lines in their spectra, thereby providing the basis for the Henry Draper Catalogue of stellar spectra, published in 1901, and Henrietta Leavitt (1868–1921), who linked the brightness of Cepheid variable stars with the length of their cycles of variation, thereby enabling their distance to be calculated. And among them certainly number such groundbreaking scientists as Albert Einstein (1879–1955), the German-born American theoretical physicist whose general theory of relativity, first propounded

The Crab Nebula is also known today as M1, after Charles Messier, who compiled a list of non-stellar astronomical objects.

Albert Einstein's general theory of relativity has helped us to understand the universe better.

in 1915 – which includes the celebrated equation $E = mc^2$ (where E stands for energy, m represents mass and c signifies the speed of light in a vacuum) – helped to explain the relationship between space and time. Einstein was also the first correctly to surmise that rays of light are bent, or deflected, by gravitational fields.

The work of other notable astronomers is commemorated by the tools and instruments that have been named in their honour. The Hertzsprung–Russell (HR) diagram, for example, which is used to track the evolution of stars, is named for Ejnar Hertzsprung (1873–1967) and Henry Russell (1877–1957), respectively a Dane and an American who both focused on the relationship between stars' brightness, or absolute magnitude, and their colour and temperature.

$$E = mc^2$$

MODERN ASTRONOMY

Apart from Stephen Hawking (b. 1942), the British theoretical physicist whose bestselling book on cosmology, *A Brief History of Time* (1988), became a publishing phenomenon, Hubble is perhaps the name currently best known in connection with astronomy. Edwin Hubble (1889–1953) was a classifier of galaxy types and the proposer, in 1929, of Hubble's law of the expanding universe (which holds that a galaxy's speed of movement from the Earth is proportional to its distance from the Earth, with the furthest galaxies moving the fastest), from which the Hubble constant, the rate at which the universe is thought to be expanding, is in turn derived. Although Hubble is a significant figure in the history of astronomy by any standards, his namesake telescope, the Hubble Space Telescope (HST), has arguably eclipsed him in fame. It was launched into space by America's National Aeronautics and Space Administration (NASA) and the European Space

Astrophysicist Stephen Hawking has contributed significantly to our knowledge of black holes.

The mighty Hubble Space Telescope was named after Edwin Hubble, who has also given his name to Hubble's law and Hubble's constant.

The whirlpool galaxy's spiral structure was first observed by the Earl of Rosse.

Agency (ESA) on the space shuttle *Discovery* in 1990. Since 1993, when a malfunction was repaired, this remote-controlled, optical telescope has beamed images of the utmost clarity and mesmerising beauty back to salivating scientists on Earth.

Powered by solar panels, the HST orbits the Earth once every 97 minutes at a rate of 27,700 km (around 17,210 miles) an hour at an altitude of around 610 km (380 miles). It measures 13.1 m (43 ft) in length and has a main mirror 2.4 m (nearly 8 ft) in diameter, unquestionably qualifying it as a large telescope.

Effective giant telescopes – which are mainly reflectors rather than refractors due to the lightness of mirrors compared to lenses – date from 1845. This was the year in which the Earl of Rosse, an Irish astronomer, discovered the spiral structure of the Whirlpool Galaxy, thanks to the 180-cm (71-in) mirror that comprised an integral part of the large telescope that he had built. Schmidt telescopes – named for their inventor, the Estonian-born German optics specialist

Bernhard Schmidt (1879–1935), who created the first in 1930 – were equally pioneering pieces of equipment. They contain a spherical mirror and corrective lens that, in conjunction, capture an image of a large area of the night sky in great detail. Today, a number of enormous telescopes situated at isolated spots around the world give astronomers extraordinarily detailed views of the universe. One of the oldest is the Hale Telescope – named for George Hale (1868–1938), the director of the American Mount Wilson Observatory – on California's Palomar Mountain, which went into operation in 1948 and boasts a 5.08-m (nearly

The massive telescopes that are ranged around the world today gather a wealth of astronomical information that is unprecedented in its detail.

17-ft) mirror. Two of the latest include the Keck I and II telescopes, the first of which began operating at the top of Hawaii's Mauna Kea mountain in 1992. Their main mirrors measure a massive 10 m (nearly 33 ft) in diameter, and because each comprises 36 computer-controlled, hexagonal segments, it does not bend under its own weight.

Radio waves are captured by radio telescopes that comprise an enormous dish and antenna, to put it simply.

A further new technique combines images from different telescopes to give the best-possible image. At the European Southern Observatory's site in Chile, four 8.2-m (27-ft) telescopes, collectively called the Very Large Telescope (VLT), together have the power of one 16.4-m (nearly 54-ft) telescope. Another crucial innovation was the altazimuth type of mount that was first used to support the USSR's 6-m (20-ft) Zelenchukskaya telescope in 1976. The altazimuth mounting has since become standard worldwide because it enables telescopes to be moved around axes both horizontally ('in azimuth') and vertically ('in altitude'), whereas the equatorial mounting that had previously prevailed is limited to observing the movement of the stars around either the northern or southern celestial pole.

Along with large telescopes of various types, professional astronomers of the 21st century have inherited, as well as developed, a number of scientific fields that enable them to learn more about the universe. Radio astronomy, for example, is a science that was pioneered by the American radio engineer Karl Jansky (1905–50) when he discovered, in 1932, that electrons in a galaxy's magnetic field emit radio waves (a form of electromagnetic radiation) powerful enough to interfere with radio signals on Earth, and by Martin Ryle (1918–84), who catalogued 5,000 radio sources. Radio waves are collected by radio telescopes, which take

A view of the Moon, captured with an ultraviolet camera during the Deep Space Program Science Experiment being carried out on the US spacecraft <u>Clementine</u>.

the form of a huge metal dish that reflects and focuses the waves onto an antenna, which in turn relays the information received to a computer for analysis. The largest radio telescope, at Arecibo, Puerto Rico, measures 305 m (1,000 ft) in diameter. It is largely due to radio astronomy that the first-known quasar, 3C 273, was discovered in 1963, the first-known pulsar, PSR 1919+21, following in 1967.

Apart from radio astronomy, other now well-established electromagnetic-radiation-based astronomical specialties include infrared astronomy (which studies infrared radiation produced by cool gases and dust), ultraviolet astronomy (which homes in on the cosmic ultraviolet emissions produced by hot stars), X-ray astronomy (which detects X-rays emitted by intensely hot gas sources) and gamma-ray astronomy (which focuses on high-frequency electromagnetic radiation and its sources, such as pulsars and quasars). All of these have added significant pieces to the cosmological jigsaw puzzle that we are slowly completing. At the heart of these analytical branches of astronomy lies spectroscopy, or the analysis of a spectrum of electromagnetic radiation – ranging from long radio waves and microwaves through infrared, visible-

Kepler's supernova remnant: a view from the Chandra X-ray Observatory.

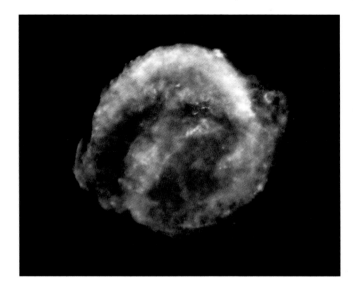

light rays, ultraviolet radiation and X-rays to short gamma rays – to determine its properties. By breaking down this barcode-like spectrum into its component parts, astronomers can learn more about its source. The first spectroscope, or spectrometer, is credited to the German lens-maker Joseph von Fraunhofer (1787–1836) in 1814. Today, spectrographs working in conjunction with computers, electronic cameras containing charge-coupled devices (CCDs) and green, blue and red filters perform a crucial role in splitting electromagnetic radiation, including starlight, into wavelengths and analysing its components.

There are observatories dedicated to these branches of astronomy, such as those housing the UK Infrared Telescope and the NASA Infrared Telescope Facility, both of which resemble optical telescopes and are based on Hawaii. Nor are they Earth-bound any longer, for satellites are now used to scan space in search of data. The mission of the *Uhuru* satellite, which was launched in 1970, for instance, was to map X-ray wavelengths in space, while the Infra-red Astronomical Satellite (IRAS) identified five new comets, among other marvels, in 1983, and the Goddard High Resolution Spectrograph continues its work aboard the HST.

This composite image compares a visible-light picture of the glowing Trifid Nebula (left panel) with infrared views from NASA's Spitzer Space Telescope (the remaining three panels). The Trifid Nebula is a giant, star-forming cloud of gas and dust 5,400 light years away from Earth in the constellation of Sagittarius.

The Soviet cosmonaut Yuri Gagarin was the first person in space to be launched into space, in 1961.

Neil Armstrong was the first man to set foot on the Moon, in 1969.

Then pioneering instruments, such specialist telescopes and their host observatories in space – notably NASA's HST, the Compton Gamma-ray Observatory (which operated from between 1991 and 2000), the Chandra X-ray Observatory (launched in 1999) and the Spitzer Space Telescope (launched in 2003) – are now almost commonplace

That the sum of the world's astronomical knowledge advanced by leaps and bounds during the second half of the 20th century is largely due to the breaching of the Earth's atmosphere and the opening-up of space to rockets, satellites and spacecraft. Space exploration was first envisaged by the Russian rocket theorist Konstantin Tsiolkovsky (1857–1935), the author of the visionary book *Exploration of Cosmic Space by Means of Reaction Devices* (1903); the American rocket engineer Robert Goddard (1882–1945), who built and launched the first liquid-fuelled rocket in 1926; and the German aeronautic pioneer Hermann Oberth (1894–1989), the author of *The Rocket into Interplanetary Space* (1923)

A NASA Space Shuttle.

The Soviet space station Mir was launched into space in 1986.

and *The Road to Space Travel* (1929). World War II would see their dreams become a deadly reality, with Wernher von Braun (1912–77) developing Germany's V–2, a rocket-powered, ballistic missile whose primary target was London, England. Thereafter, the superpower rivalry between the USA and the Soviet Union during the Cold War fuelled the 'space race' that resulted in von Braun (now on America's side) designing the *Redstone* and *Saturn* rockets that first launched Americans into space. In the Soviet Union, it was Sergei Korolev (1906–66) who was responsible for producing his nation's first intercontinental missile, followed by the *Sputnik 1* satellite (which was launched in 1957, beating the USA's *Explorer 1* satellite into space by a mere four months) and the *Vostok*, *Voskhod* and *Soyuz* spacecraft. While the Soviet cosmonaut Yuri Gagarin was the first individual to orbit the Earth (in 1961), the American astronaut Neil Armstrong was the first man on the Moon (in 1969). And thanks to these satellites, probes, shuttles, space stations and their onboard cameras, from the late 1950s, humans have been treated to ever-clearer views of the alien landscapes of other planets, starting with Venus in 1975.

An artist's representation illustrates the positions of the _Voyager_ spacecraft in relation to structures formed around the Sun by the solar wind. Also illustrated is the termination shock, a violent region that the spacecraft must pass through before reaching the outer limits of the solar system. At the termination shock, the supersonic solar wind abruptly slows from an average speed of 400 km per second to less than 100 km per second (900,000 to less than 225,000 miles per hour). Beyond the termination shock is the solar system's final frontier, the heliosheath, a vast region where the turbulent and hot solar wind is compressed as it presses outwards against the interstellar wind beyond the heliopause.

REACH FOR THE STARS!

If space, to borrow a celebrated phrase from the early space-age television series *Star Trek*, is the final frontier, the boundaries that have restricted our knowledge of the universe are now being pushed back at an unprecedented rate almost daily. The sky is no longer the limit for 21st-century astronomers, and in response to their constant probing, the cosmos is slowly yielding its secrets. Embark on your own personal voyage of discovery, and you are guaranteed a fascinating ride around the universe, whether you decide to limit yourself to a whistle-stop sightseeing tour conducted at the speed of light or feel tempted to slow down in order take a more lingering look. And once you reach for the stars, who knows where your journey will end? It may just be that in future years the fruits of your own observations will earn you a mention in an astronomy handbook like this one!

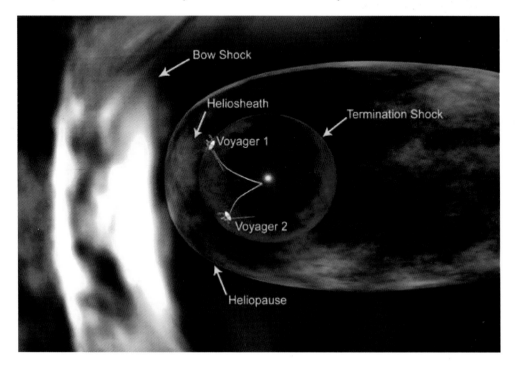

The universe is thought to have come into being when a singularity exploded.

In the Beginning

Before blasting off on a voyage of discovery around the solar system, and thence past the Oort Cloud into the Milky Way and beyond, first entering the domain of the Local Group (of galaxies), and then that of the Local Supercluster (again of galaxies), before finally approaching the very edges of the universe, we should briefly dwell on cosmogonical matters. We know that the universe contains many wonders, and are learning more about them all the

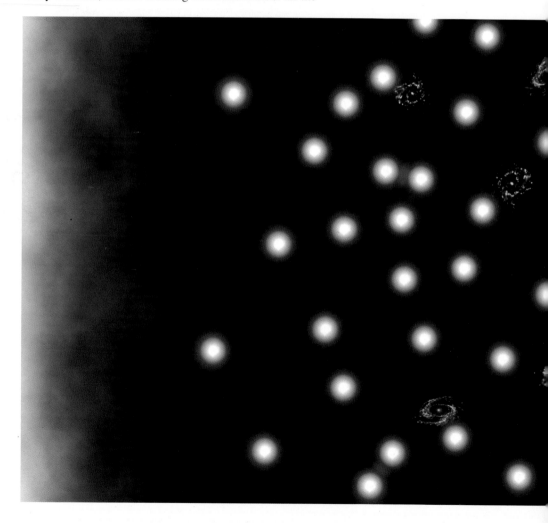

time. But how did they come into being? Come to think of it, how did the universe itself come into being?

Some Christians remain convinced that the universe, or cosmos, was created by God literally in accordance with the acts of creation described in the Old Testament book of Genesis (see page 15). There are also cosmologists – scientists whose area of expertise is the evolution and structure of the universe – who believe that a divine creator is not incompatible with a scientific explanation of the formation of the cosmos and its components. Without making any judgements on what is regrettably often a bitterly contentious issue, let us simply review the facts, always bearing in mind the possibility that future discoveries may force us to reassess our early 21st-century viewpoint.

THE BIG-BANG THEORY

It was largely thanks to new technical capabilities, which prompted an increasing focus on studying electromagnetic radiation in all of its forms during the second half of the 20th century, that astronomers were able to construct an ever-more detailed picture of the universe. Armed with this knowledge, many cosmologists have advanced theories to explain how the universe came into being, typically based on the acceptance that it is expanding. Today, the big-bang theory – built on the work of such groundbreaking astronomers as Vesto

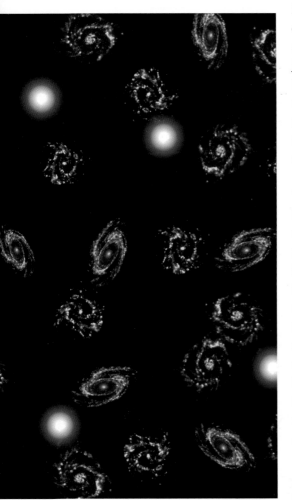

As this image of 'baby' galaxies shows, the universe is still expanding.

Melvin Slipher (1875–1969), who calculated the speed at which 25 galaxies were moving, and his American compatriot Edwin Hubble (see page 31) – is regarded as providing the most persuasive explanation, although it remains both incomplete and unproven.

Georges Lemaître (1894–1966), a Belgian scientist who, in 1931, suggested that the universe was created when a 'primeval atom' exploded in space, is credited with being the originator of the big-bang theory. Notable cosmologists such as George Gamow (1904–68) and Alan Guth (b. 1947), both US citizens, subsequently refined it. Crucially, in 1965, US scientists Arno Penzias (b. 1933) and Robert Wilson (b. 1936) detected a weak radio signal reaching Earth from every direction, produced by a source whose temperature was –270°C (absolute zero), prompting them to conclude that this must be evidence of the (microwave) background radiation caused by the cooling of the heat generated by the big bang.

In extremely simplified terms, the big-bang theory states that around 12,000 to 15,000 million (or 12 to 15 billion) years ago – yet before time existed – an incredibly hot and dense point in space (termed 'a singularity') suddenly exploded with unimaginable force, maybe in response to an unknown trigger or perhaps simply spontaneously. As well as initiating time, the energy (or electromagnetic radiation)

The universe is thought to have come into being when a singularity exploded.

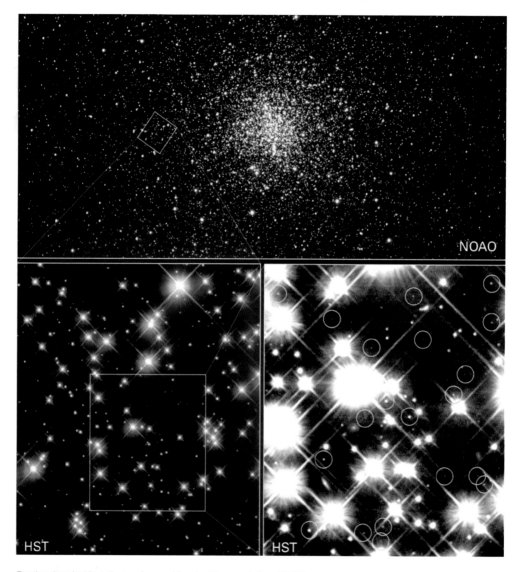

Peering deep inside a cluster of several hundred thousand stars, NASA's Hubble Space Telescope has uncovered the oldest, burned-out stars in our Milky Way galaxy, giving astronomers a fresh reading on the age of the universe. Located in the globular cluster M4, these small, burned-out stars – called white dwarfs – are about 12 to 13 billion years old. By adding the 1 billion years that it took the cluster to form after the big bang, astronomers found that the age of the white dwarfs agrees with previous estimates that the universe is 13 to 14 billion years old.

generated by this 'big bang' created equal amounts of matter and antimatter, which initially annihilated one another. As the universe inflated, the balance gradually shifted in favour of matter, however, particles of which began to adhere to one another to create the nuclei of the first three elements, hydrogen, helium and lithium, as well as electrons and particles of dark matter. Thereafter (and the initial process apparently took only three minutes), radiation and matter decoupled, and as the particles of gaseous matter slowed down and cooled, they began to form stars and galaxies, which continue to evolve as they travel away from one another. According to the latest calculation of Hubble's constant (see page 31), which was performed using data

HOW DO WE KNOW THAT THE UNIVERSE IS EXPANDING?

We know that the universe is expanding because we know that galaxies, including our own galaxy, the Milky Way, are moving away from one another. And we know that they are doing this because when a galaxy's spectrum is examined with a spectrometer, spectral lines are visible towards the red end of the spectrum, where the frequency is decreased and the wavelengths lengthen, signifying that the galaxy is moving away from us. This redshift is the result of the Doppler effect, or change in apparent wavelength, or frequency of a wave, due to the relative motion between the source and the observer, a phenomenon that was named for its discoverer, the Austrian physicist Christian Doppler (1803–53). (That said, the latest cosmological thinking holds that the cosmological redshift mimics, but is not caused by, the Doppler effect.) The greater the number of lines that are redshifted, the faster a galaxy is moving, and the further away it is from us.

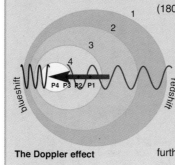

The Doppler effect

Dark matter is the mysterious, invisible material that lies between galaxies.

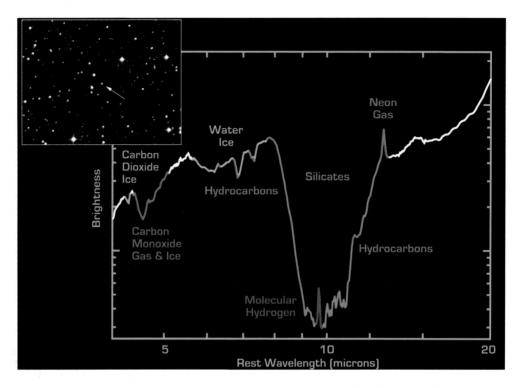

NASA's Spitzer Space Telescope has detected the building blocks of life in the distant universe, albeit in a violent milieu. Training its powerful, infrared eye on a faint object located at a distance of 3.2 billion light years, Spitzer has observed the presence of water and organic molecules in the galaxy IRAS F00183-7111. With an active galactic nucleus, this is one of the most luminous galaxies in the universe, rivalling the energy output of a quasar. Because it is heavily obscured by dust (see the visible-light image in the inset), most of its luminosity is radiated at infrared wavelengths.

The infrared spectrograph instrument on board Spitzer breaks light into its constituent colours, much as a prism does for visible light. The image shows a low-resolution spectrum of the galaxy obtained by the spectrograph at wavelengths between 4 and 20 microns. Spectra are graphical representations of a celestial object's unique blend of light. Characteristic patterns, or fingerprints, within the spectra allow astronomers to identify the object's chemical composition and to determine such physical properties as temperature and density.

gathered by the Hubble Space Telescope, the universe is now expanding at a rate of 73 km (around 45 miles) per second per megaparsec (or million parsecs) distance.

DARK MATTER AND UNIVERSAL INFLUENCES

The universe contains a myriad of objects, including galaxies, planets and stars, many of which we will look at in greater detail in later chapters. But what lies between them? Scientists call this mysterious, invisible material 'dark matter', and estimate that more than 90 per cent of the universe's mass consists of it. We cannot see it, but we can detect the effect that its gravity has in stabilising the positions of many of the cosmos' visible components. Apart from gravity, or the gravitational force, three other forces exert their influences over the universe: the electromagnetic force, which, as its name suggests, influences electricity and magnetism; the strong force, which keeps atoms' nuclei together; and the weak force, which controls how stars shine.

WILL IT COME TO THE BIG CRUNCH?

The universe is without doubt evolving. But cosmologists do not know for sure whether it will exist forever, or whether it will some day come to an end, be it as the result of a dramatic 'big crunch' or by fading slowly away. According to the big-crunch theory, gravity is slowing the expansion of the universe. If the density of the gravity-producing matter within it is greater than a critical density, once this 'closed universe' has reached the limits of its expansion, it will contract. This will cause the matter to become compacted and to heat up until it eventually breaks down into subatomic particles. It will then be consumed by a super-supermassive black hole. Crunch!

In 1997, however, it was calculated that the universe's mass did not exceed critical density. In 1998, astronomers studying type 1 supernovae in distant galaxies concluded that the universe's rate of expansion is accelerating rather than decelerating. As a result, the big crunch seems an increasingly unlikely cosmic conclusion. An alternative possibility is that if the universe is low in mass, the gas and dust that form and fuel stars will run out one day. The ultimate scenario is again a single, rapacious black hole that may itself ultimately explode in a burst of radiation, leaving this 'open universe' a freezing, dark expanse of emptiness.

Ultraviolet image of NGC 5128 (Centaurus-A). This unusual galaxy is believed to be the result of a collision of two normal galaxies. The blue regions toward the top are thought to be areas of star formation induced by powerful jets originating from a central black hole.

The Solar System

The solar system to which our planet, the Earth, belongs is believed to have come into being around 4.6 billion years ago. According to the theory first proposed by French astronomer Pierre de Laplace in 1796 (which has since become generally accepted, albeit with some modifications), its formation was kick-started when a shockwave emanating from a supernova triggered a giant molecular cloud (GMC) – a massive cloud of hydrogen, helium, rock, metal, dust and snow spinning around in space – to contract and the Sun, the star around which the planets revolve, to be born at its centre.

It was not long – in space time, at least – before the planets themselves (apart from Pluto) slowly started to take shape, as particles of the material within the solar nebula, or protoplanetary disc (see page 191), cocooning the young Sun began to clump together to form planetesimals, or rocky objects whose increasing mass gave them gravity, which in turn attracted more matter, including other planetesimals, to them. Eventually, these core planetesimals became large enough to be called protoplanets, which similarly aggregated and grew larger.

Fast-forwarding to around 1 billion years later, the process of accumulation and evolution resulted in four 'terrestrial' planets with a molten-metal core and rocky mantle and crust. Mercury, Venus, the Earth and Mars, or the inner planets, are each held in place by the Sun's gravity and follow individual, elliptical orbits around it. Around 4 billion years ago, these inner planets endured the Late Heavy Bombardment, when they were repeatedly struck by a barrage of interplanetary debris, evidence of which survives as craters.

Separated from the inner planets by the Asteroid Belt are the five outer planets: the gas giants Jupiter, Saturn, Uranus and Neptune – all of which have rings and moons – and then Pluto, a sort of runt of the planetary litter in that this tiny planet was created from leftover materials and has no rings

The planets in the solar system are Mercury, Venus, the Earth, Mars, Jupiter, Saturn, Uranus, Neptune and Pluto.

encircling it. Indeed, many astronomers, particularly since the discovery of Sedna, a similar celestial object, believe that Pluto is too small, and its orbit too eccentric, to merit the title 'planet', and that it should instead be classified as a Kuiper Belt object.

If the planets are the major members of the solar-system club, the minor members include such solar-nebula debris as meteoroids; the asteroids (which are also known as minor planets or planetoids) that mainly occupy the Asteroid Belt; the mysterious objects that inhabit the Kuiper Belt, which lies between Neptune's orbit and the Oort Cloud; and the comets that orbit the Sun beyond Pluto, within the confines of the Oort Cloud.

Let us now take a tour of our solar system, starting with the Sun blazing away at its centre, and then travelling outwards, pausing to examine each planet in turn as it slowly spins on its tilted axis while orbiting the Sun, and in the process discovering that each grows successively colder and takes longer to complete a full orbit the further away from the Sun it lies. On our whistle-stop tour, we'll also be taking in our Moon and its many natural-satellite cousins, as well as the minor members of the solar system. (And if you come across any terms that aren't explained fully, or that you don't understand, on our journey, refer to the glossary on page 234 onwards.)

This colour image of the Sun, Earth and Venus was taken by the Voyager 1 spacecraft on 14 February 1990, when it was approximately 32 degrees above the plane of the ecliptic and at a slant-range distance of approximately 6.4 billion km (4 billion miles). It is the first, and maybe the only time, that we will ever see our solar system from such a vantage point.

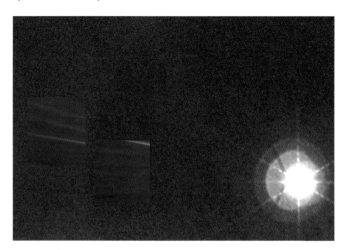

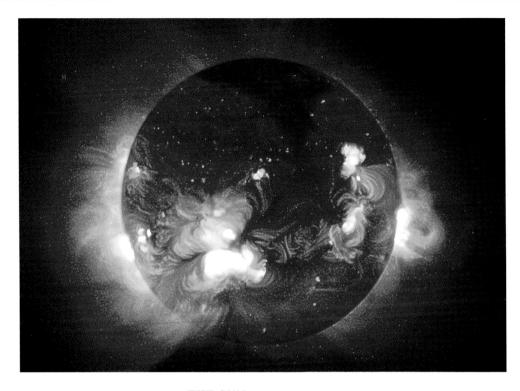

An X-ray image of the Sun.

THE SUN

Diameter: 1.4 million km
(869,944 miles)
Mean core temperature:
15 million°C (27 million°F)
Mean surface temperature:
5,500°C (9,932°F)
**Rotational period at its
equator:** 25.4 days
Visible magnitude: –27

THE SUN

If you've ever suffered from sunburn, you'll be all too well
aware of the intensity of the energy generated by the Sun.
Now halfway through its lifespan, the Sun is currently a
main-sequence star (and specifically, a yellow dwarf of the
spectral type G2, see page 136). Its luminosity is estimated at
390 quintillion megawatts and its energy is generated by the
nuclear fusion that takes place at its fiery core, where the
temperature is a mind-boggling 15 million°C (27 million°F).
It is in this inferno that atoms of hydrogen gas shatter, lose
their electrons and break down into nuclei. And in these
conditions, when the protons in four hydrogen nuclei come
together, they fuse to make a helium nucleus. (This process
of nuclear fusion also releases positrons, neutrinos and
gamma-ray photons.) At its core, the Sun comprises around

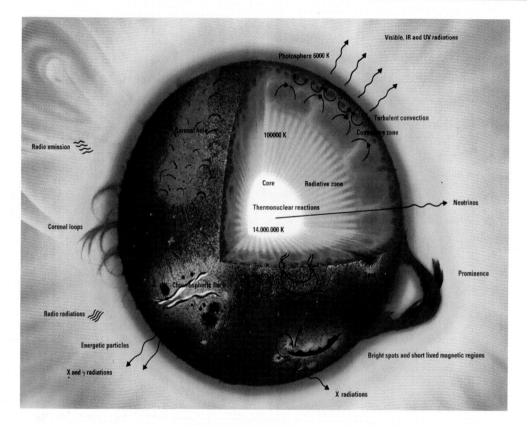

Visible, IR and UV radiations

Photosphere 6000 K

Turbulent convection

Coronal hole

100000 K

Convective zone

Radio emission

Core   Radiative zone

Neutrinos

Thermonuclear reactions

14.000.000 K

Coronal loops

Prominence

Chromospheric flare

Radio radiations

Energetic particles

Bright spots and short lived magnetic regions

X and γ radiations

X radiations

A cross-section of the Sun.

34 per cent hydrogen and about 64 per cent helium. When the Sun's supply of hydrogen is exhausted, in roughly 5 billion years' time, it is thought that it will expand in diameter to become about 30 times its present size, while its luminosity will increase a thousand-fold; in short, it will become a red giant, or giant star, then a dying white dwarf and ultimately a stellar ghost, or black dwarf. (See pages 145–8 for details on stellar evolution and what is likely to happen to the Sun.)

The photosphere and sunspots

From the Sun's core, the energy created by nuclear fusion passes through a region called the radiative zone as photons, which then enter the convective zone and are carried by convection cells – sorts of pulses, waves and eddies of hot gas – to the photosphere, or Sun's surface.

A 500-km- (311-mile-) thick ocean of gas, the photosphere's temperature ranges from between 8,500°C (15,332°F) nearest the core to 4,200°C (7,592°F) nearest the chromosphere. That said, there are regions of the photosphere where the temperature drops by about 1,500°C (2,732°F) below average. These regions, or sunspots, appear dark when viewed from Earth. Sunspots are thought to form in so-called 'active regions' within the Sun's magnetic field, each 11-year-long cycle of appearance starting near the poles and spreading to the solar equator before eventually dying down again. Why does this happen? Well, the Sun rotates, but because it is a ball of gas rather than a solid sphere, it does not do so evenly, which means that its equator takes around 25.4 days to make a full turn, while its poles take 35 days. In addition, waves of sound created in the convection zone cause the photosphere to oscillate, or swing from side to side. As a result, parts of its magnetic field become twisted and break through the surface as loops, with a north magnetic pole at one end and a south magnetic pole at the other, that repel any hot gas attempting to rise from the core, creating a sunspot. In the anatomy of a sunspot, which may have a lifespan of hours or weeks, the umbra is the relatively cool centre, its hotter surroundings being termed the penumbra. The smallest sunspots are termed pores, and the largest are clusters called sunspot groups.

Sunspots are just one of a number of types of solar events that occur on the

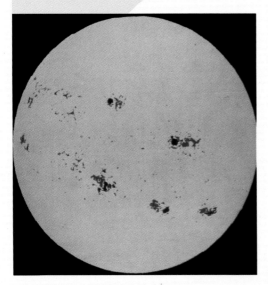

Sunspots appear dark from Earth.

Sun's surface. Others include faculae (singular 'facula'), or hot, white areas that precede and follow sunspots; granulation, or the mottled appearance caused by convection cells; and the magnetic carpet, or loops of magnetism that give the photosphere a rug-like look. (Warning! If you want to see any type of solar event for yourself, remember never to look directly at the Sun, and to take the precautions outlined on page 208.)

The chromosphere and corona

Having passed through the photosphere, the photons of solar energy have to negotiate the Sun's atmosphere, which consists of the chromosphere and the corona.

The chromosphere is believed to be about 5,000 km (3,107 miles) thick, and to have a temperature range of between 4,000°C (7,232°F), where it abuts the photosphere, and 500,000°C (900,032°F), where it merges with the corona. With the atmosphere comprising 73 per cent hydrogen and 25 per cent helium, the ratio between the two gases here is dramatically different from that in the core. Solar events occur in the chromosphere, too, such as spicules, or jets of gas that shoot into the corona, and their larger brethren, macrospicules; flares, or explosions that erupt above sunspot groups; and prominences, or clouds and filaments of gas, which are situated above sunspots and stretch from the chromosphere into the corona.

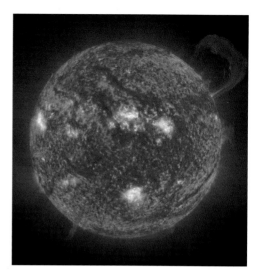

An Extreme Ultraviolet Imaging Telescope (EIT) image of a huge, handle-shaped prominence taken on 14 September 1999. Prominences are huge clouds of relatively cool, dense plasma suspended in the Sun's hot, thin corona. At times, they can erupt, escaping the Sun's atmosphere. The hottest areas appear almost white, while the darker-red areas indicate cooler temperatures.

A solar flare.

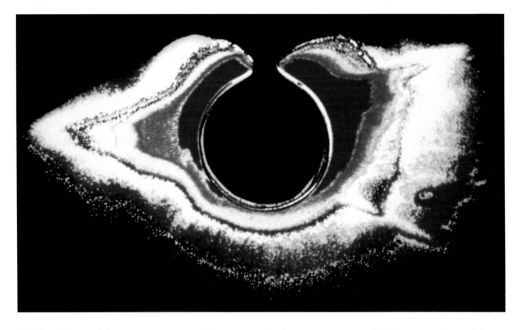

An infrared image of the Sun's corona.

The corona is the uppermost and thinnest portion of the solar atmosphere. Perhaps surprisingly, however, it is hotter than the chromosphere, for the temperature within it can rise to 3 million°C (5,400,032°F) and higher. The corona's outer reaches are the solar wind's point of departure into space: from here, it carries charged particles (mainly protons and electrons) through the heliosphere, an area that extends up to 15 billion km (9.3 billion miles) from the Sun, at speeds that vary according to the violence of any solar events that may or may not be taking place on the Sun.

Once out of the Sun's atmosphere, solar energy mainly takes the form of visible light and infrared radiation, and it is this that, on the one hand, lights up the Earth by day following its arrival 8.3 minutes after leaving the Sun, and, on the other, can cause sunburn.

A typical view of the Sun's rays.

PROFESSIONAL SOLAR OBSERVATIONS

Professional astronomers use a number of different instruments with which to observe the Sun and collect and analyse solar-generated data. The following are just a few of them.

● Spectographs are used to analyse the light emitted by the photosphere in order to determine its make-up.

● Magnetographs produce magnetograms, or maps of the Sun's magnetic fields, for analysis.

● Solar, or optical, telescopes comprise a moving mirror (a heliostat) atop a tower (which acts as the telescope's 'body'), at the base of which are instruments of analysis; many are primarily concerned with solar events. The USA's McMath-Pierce Solar Telescope, for instance, focuses on sunspots.

● Radio telescopes, including Japan's Nobeyama Radioheliograph, collect data from the Sun's active regions.

● Neutrino telescopes, such as the Sudbury Neutrino Observatory, which is situated 2 km (1¼ miles) underground in Ontario, Canada, detect the number of solar neutrinos that reach the Earth from the Sun.

● Spacecraft like NASA's TRACE (Transition Region and Coronal Explorer), which was launched in 1998, study the Sun at closer quarters. TRACE concentrates on the transitional region between the chromosphere and corona, as well as on the corona itself.

This composite image combines Extreme Ultraviolet Imaging Telescope (EIT) images from three wavelengths (171, 195 and 284 angstrom) into one that reveals solar features unique to each wavelength. Since the EIT images come to us from the spacecraft in black and white, they are colour-coded for easy identification. For this image, the nearly simultaneous images from May 1998 were each given a colour code (red, yellow and blue) and merged into one.

The solar eclipse of 16 February 1980.

Solar eclipses

Although the Sun's photosphere is seen as the solar disc from the Earth, the chromosphere and corona are only visible during a total eclipse, when the Moon passes directly in front of the Sun, thereby blocking it from a particular Earthly viewpoint. The centre of the Moon's shadow as it falls on the Earth during an eclipse is the umbra, while the lighter ring, or half-shadow, that surrounds the umbra is the penumbra. During a total solar eclipse, when the umbra bathes observers on the Earth in darkness, the corona is apparent against the lowering sky as a pearly-white halo, sometimes displaying wispy projections that encircle the chromosphere, which in turn typically manifests itself as a pinkish-red ring around the black lunar disc. Watching the Moon 'eating up' the Sun as it moves across it from west to east is an unnerving sight, made even more unearthly as totality approaches by the backdrop of twinkling stars. For a few minutes, day becomes night, the air rapidly cools and birds fall silent, adding to the eerie effect. Small wonder that people once thought that the end of the world was nigh!

Other types of solar eclipses include partial eclipses and annular eclipses. A solar eclipse is classified as a partial eclipse when the Moon passes in front of the Sun from the vantage point of an Earthling who is in the Moon's penumbral shadow, giving him or her the impression that the Moon has bitten a chunk out of the Sun. An annular eclipse

The sudden onset of darkness when a solar eclipse occurs by day can feel very eerie!

DATES FOR YOUR DIARY

Although between two and five solar eclipses are visible from the Earth most years, you may rarely witness a total solar eclipse. This type of solar eclipse generally lasts for anything from a few seconds to seven-and-a-half minutes, depending on how close the Moon is to Earth. Blink, or hastily complete a task, and you could miss it, which is why it is important to be ready and waiting. Total solar eclipses are scheduled to occur on the following dates in the near future:

- **29 March 2006,** when the eclipse will be visible from the Atlantic through northern Africa to central Asia;
- **1 August 2008,** when people in the Arctic, Greenland, northern Canada, Siberia and China will be able to observe the eclipse;
- **22 July 2009,** when the eclipse will be visible from India, China and the Pacific region;
- **11 July 2010,** when observers in the South Pacific will be able to see the eclipse.

If you are ever lucky enough to watch a total solar eclipse, before and after totality you may be treated to the sight of the phenomena called Baily's Beads (a string of dazzling lights) and the Diamond Ring (a single, brilliant light), both of which are caused by light from the Sun's photosphere escaping between the jagged peaks of the lunar mountains that lie around the edge of the Moon's disc.

occurs when the Moon is at its apogee, or at its most distant point from the Earth, when it appears too small to blot out the Sun and the outer regions of the photosphere (the annulus) blaze out around the dark, lunar disc like a fiery hoop.

(See page 208 for information on how to observe any solar eclipse without damaging your eyesight.)

An annular eclipse.

Keeping the planets in line

Despite being made of gas, the Sun's mass is roughly 745 times greater than that of all of the planets put together, which is why it exerts such a powerful gravitational pull on them. Unable to break free, they must continue to follow their elliptical orbits around the Sun as though they were connected to it by the strongest of chains, the planetary series starting with Mercury and finishing with Pluto (or perhaps with Neptune or the newly discovered Sedna).

The planets cannot break free of the Sun's gravitational pull.

THE PLANETS AND THEIR NATURAL SATELLITES

Mercury

Mercury was named after the Roman messenger god. His Greek counterpart was Hermes, a fleet-footed deity whose winged sandals and helmet enabled him to fly around the world at lightning speed. It is therefore appropriate that Mercury should orbit the Sun faster than any other planet, i.e., in 88 days, although the rate at which it completes a rotation while spinning nearly vertically on its axis is a relatively slow 59 days.

An inner, inferior, terrestrial planet, two-thirds of Mercury's mass is thought to consist of an iron core that generates a weak magnetic field. Its mantle and crust are

MERCURY

Diameter: 4,879 km
(3,032 miles)
**Average distance from
the Sun:** 58 million km
(36 million miles)
Mean surface temperature:
167°C (333°F)
Orbital period: 88 days
Number of moons: 0
Visible magnitude: −1.9

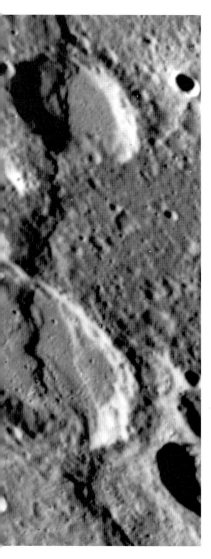

Source: Mariner 10, March 1975. As can be seen in this image, the surface of Mercury is heavily cratered. The prominent scarp that snakes up the image was named Discovery Rupes. This feature is thought to have been formed as the planet compressed, possibly due to the cooling of the planet.

made up of silicate rock, and it barely has an atmosphere or gravity. Because it lies so close to the Sun, Mercury bears the full brunt of the blistering rays of solar radiation. Nevertheless, it has cooled and shrunk since it took shape 4.6 billion years ago, causing ridges or cliffs ('rupes', singular 'rupis') to form on its surface. In its youth, it was bombarded and battered by an onslaught of meteoroids during the Late Heavy Bombardment, so that its surface is today covered with impact craters – named for human luminaries of the arts, a stellar A to Z that runs from the Arab poet Abu Nuwas (*c.* 1756–1810) to the French novelist Emile Zola (1840–1902) – surrounded by rocky areas and larval plains ('planitia', singular 'planities'). Other notable features of Mercury's Moon-like landscape include mountain chains ('montes', singular 'mons') and the massive Caloris Basin, a lowland plain created when an asteroid smashed into the planet.

Observing Mercury and transits of Mercury

Be warned that Mercury, which, when seen from the Earth, is positioned a little above or below the Sun, is not the easiest of planets to observe. Your best chances of seeing it are just after the Sun has set at the start of spring or just before the Sun rises in early autumn, when you should scan the horizon looking for a bright, yellowish star. You may be able to glimpse it unaided, but will have a better view if you use binoculars. (Wait until the Sun has sunk below the horizon before using them.) Better still, if you have a reasonably powerful telescope, you may be able to confirm that Mercury displays phases (see page 79), and may even see faint evidence of the planet's craters, plains, mountains and ridges.

Every few years or so, when the Sun, Mercury and Earth are in alignment, as viewed from an Earthly vantage point, Mercury will appear as a small, black dot moving across the face of the Sun, a phenomenon known as a transit of Mercury. Transits of Mercury are due to occur on 8 November 2006 and 9 May 2016. Before viewing one, consult page 208 to read about the precautions that you must take to safeguard your eyesight.

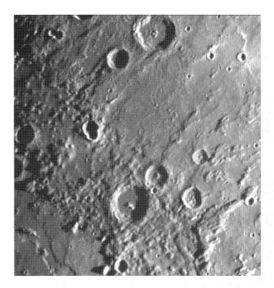

This image of the north-eastern quadrant of the Caloris Basin shows the smooth hills and domes between the inner and outer scarps and the well-developed radial system east of the outer scarp. This image was taken during Mariner 10's exploration in March 1974.

PROFESSIONAL MERCURIAL OBSERVATIONS

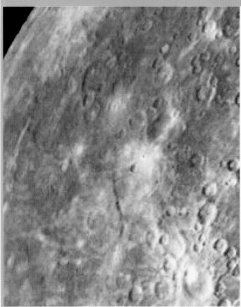

An enhanced colour image taken by Mariner 10.

Mercury's proximity to the Sun means that it is radar, rather than any optical instrument, that is the professional astronomer's friend when studying this planet. Astronomers worked out the length of Mercury's rotational period with the help of radar in 1965, and it is still used to gather information about the planet's features.

Space probes have also provided invaluable Mercurial insights. The US space probe Mariner 10 flew past Mercury on three occasions between 1974 and 1975, for instance, photographing the planet's surface and in the process discovering its magnetic field.

Venus

Venus is named after the Roman goddess of love, a deity whose beauty outshone that of all others and who was known as Aphrodite in Greece. This inner, inferior planet orbits the Sun in 225 days (a Venusian or Cyntherean year) and takes only a little longer – 243 days – to rotate once on its axis. It does this at an angle of 2.7°, and in a different direction to most of the other planets, so that it is said to have a retrograde motion. Venus has a weak magnetic field, but its atmospheric pressure is 90 times higher than that of the Earth.

Although Venus is similar in size and mass to the Earth, and is similarly a terrestrial planet, the Venusian world is very different from ours. It is the hottest planet in the solar system for a start, thanks to its extremely thick atmosphere, which

VENUS
Diameter: 12,076 km (7,504 miles)
Average distance from the Sun: 107 million km (67 million miles)
Mean surface temperature: 464°C (867°F)
Orbital period: 225 days
Number of moons: 0
Visible magnitude: −4.4

The Venusian
landscape.

mainly comprises carbon dioxide (CO_2), but with some
sulphur particles and sulphuric acid also present. Only about a
quarter of the light that emanates from the Sun penetrates the
overcast Venusian atmosphere. Infrared radiation is generated
when the rays reach the planet's rocky, yet relatively smooth,
surface, producing heat that then remains trapped beneath the
dense, fast-moving clouds. A rocky mantle surrounds Venus'
iron-and-nickel core, which is in turn enclosed by a basalt
crust, but not a very thick one because Venus is a world of
active volcanoes ('montes', singlar 'mons', which is also Latin
for 'mountain') – thousands of them! On the surface, low-
lying volcanic plains ('planitia' and 'plani', singular 'planus')
that exhibit shield-, spider-, pancake-dome- and dome-shaped
volcanoes and lava flows, as well as volcanic and meteoroid-

impact craters, predominate. There are also some highland regions ('regionis', singular 'regio', or 'terrae', singular 'terra'), including plateaux ('tesserae', singular 'tessera') that were formed when Venus' crust was pushed upwards and outwards by the heaving, seething, molten material beneath. Appropriately, most of these surface features are named after the divine Venus' fellow goddesses, so that you will find a Maat Mons, a Lavinia Planitia, a Lakshmi Planum, an Aphrodite Terra and a Tellus Tessera on a map of Venus, for example.

Observing Venus and transits of Venus
That Venus' alternative names are the Morning Star and the Evening Star gives you a good indication of the best times to see this planet. Its dense atmosphere reflects sunlight, which is why it is one of the most dazzling objects in the sky. It is brightest at its furthest elongations, when it is clearly visible with the naked eye just before sunrise and just after sunset. Venus displays phases (see page 79), which you can see if you train a pair of binoculars or a telescope on it on different nights. Note, however, that we can never view the 'full Venus', as it were, for when its sunlit disc is directed towards us in its entirety, it is obscured by the Sun. And not even binoculars or a telescope will enable you to see past the clouds that shroud the planet's surface.

Like Mercury, Venus is an inferior planet, which means that when the Sun, Venus and the Earth are in alignment, it can be seen in silhouette when viewed from the Earth as it travels across the solar disc. This is an extremely rare event, however: indeed, between 6 December 1882 and 8 June 2004, there were no transits of Venus. They take place at intervals of 8, 121.5, 8 and 105.5 years (the next will be on 6 June 2012, but will not be visible from Britain). If you are lucky enough to have one occur during your lifetime, make sure that you have taken the necessary steps to protect your eyes from the Sun before watching it (see page 208).

PROFESSIONAL VENUSIAN OBSERVATIONS

Optical instruments are almost useless to professional astronomers when it comes to observing Venus, but radar can penetrate the planet's murky atmosphere, while spacecraft have started to unlock some of its secrets.

Mariner 2 was the world's first successful interplanetary spacecraft. Launched on 27 August 1962, on an Atlas-Agena rocket, Mariner 2 passed within about 34,000 km (21,000 miles) of Venus, sending back valuable new information about interplanetary space and the Venusian atmosphere.

When it flew past Venus in 1962, for instance, the USA's Mariner 2 detected carbon dioxide in the Venusian atmosphere. The Soviet Union's Venera orbiters and landers, which specifically targeted Venus, had some notable successes, such as when Venera 7 landed on its surface in 1970; when Venera 9 sent back the first picture of its rocky landscape in 1975; when Veneras 13 and 14 took and analysed soil samples in 1982, also beaming back the first colour pictures from the surface; and when Veneras 15 and 16 orbited the planet in 1983, mapping it by radar. The USA's Pioneer-Venus orbiter had been the first to survey Venus in this way, in 1978, and the latest such mission was undertaken by the US orbiter Magellan between 1990 and 1994; it succeeded in documenting Venus' volcanic surface in unprecedented detail before burning up in its hostile atmosphere.

An image from the Magellan mission. Sapas Mons is displayed in the centre of this computer-generated, three-dimensional perspective view of the surface of Venus.

THE EARTH

Diameter: 12,728 km
(7,909 miles)
**Average distance from
the Sun:** 150 million km
(92.752 million miles)
Mean surface temperature:
15°C (59°F)
Orbital period: 365 days
Number of moons: 1

The Earth

You can't observe the Earth in the night sky, for the obvious reason that you are standing on it! For the record, it takes around 24 hours (a day) to turn on its axis, at an angle of 23.5° from the vertical. It is classified as an inner, terrestrial planet, and its thin basaltic and granitic crust covers a rocky mantle, underneath which lies an iron core whose outer region is molten, and whose centre is solid. Consisting of 77 per cent nitrogen (N), 21 per cent oxygen (O) and 2 per cent other types of gas, our atmosphere is made up of distinct layers. Rising from the Earth's surface, these are the

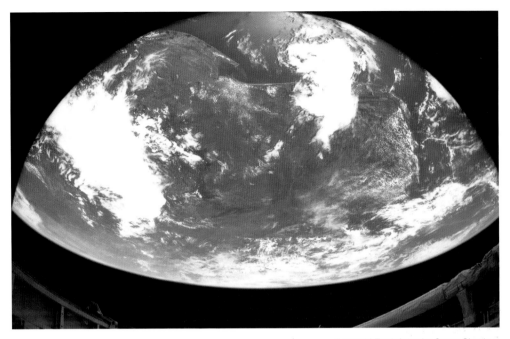

A view of Earth from the Space Shuttle.

troposphere, stratosphere, mesosphere and thermosphere (which is in turn divided into the ionosphere and the exosphere) and, finally, the all-enclosing magnetosphere, or the Earth's magnetic field.

See also pages 220-3 for more information on the changing positions of the Sun, Moon and stars in relation to the Earth during the course of our planet's year-long orbit around the Sun.

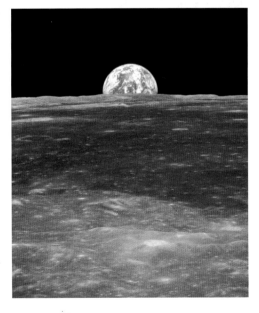

The Earth and the Moon, our planet's natural satellite.

The Moon

The Moon is not a planet, but the Earth's natural satellite. It is the closest celestial body to the Earth and orbits our planet at a rate of just over 1 km (a little over half a mile) a second. It is not certain exactly how the Moon was formed – although astronomers speculate that a collision between the Earth and a Mars-sized object created a 'big splash' of molten material that eventually evolved into the Moon – but we do know that it consists of rock, with a core that is solid at the centre and molten at the edges, and that it has no atmosphere. The Moon is a slave to the Earth's gravity, for its diameter is a quarter of that of the Earth, whose mass is also 100 times greater than the Moon's. Its period of orbit around the Earth, and the rate at which it completes one rotation on its axis (spinning at an angle of 6.7° from the vertical) are identical: 27.32 days, or a sidereal month. This means that we only ever see one side of the Moon, the near side, as the far side is perpetually hidden from us. That said, we can occasionally perceive up to 9 per cent of the far side, when the Moon oscillates as a result of its rate of orbit exceeding its speed of rotation (a phenomenon called libration). This occurs when the Moon is at its closest to us.

Like Mercury's, the surface of the Moon is pitted with craters, a testimony to the battering by meteoroids that it underwent during the Late Heavy Bombardment. Later, volcanic activity caused lava to emerge onto the lunar surface through

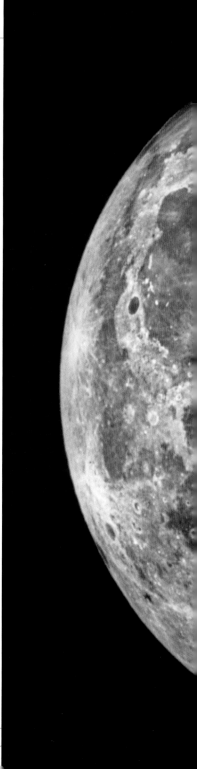

THE MOON
Diameter: 3,476 km (2,160 miles)
Average distance from the Earth: 384,400 km (238,862 miles)
Average distance from the Sun: 149 million km (93 million miles)
Mean surface temperature: –20°C (–4°F)
Orbital period around the Earth: 27.32 days
Number of moons: 0
Visible magnitude: –13

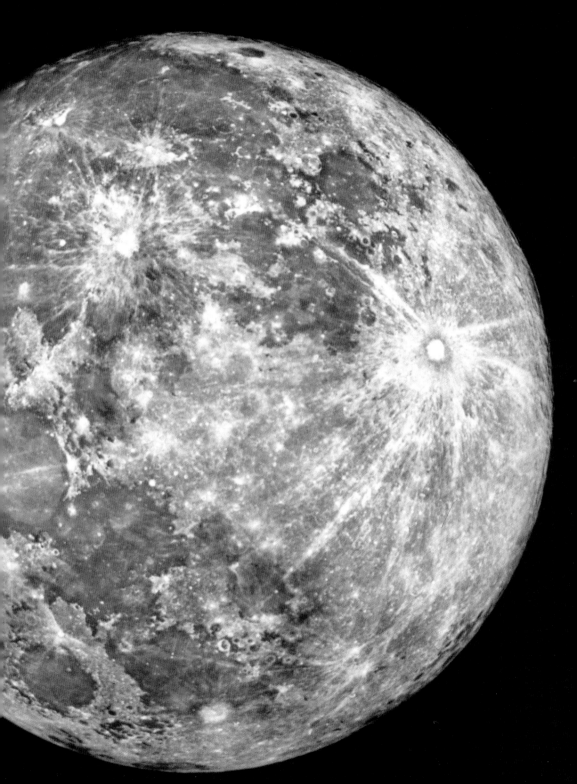

Earth as viewed from the Moon.

fissures in the crust, first flooding into lowland craters and subsequently cooling and solidifying to form the lunar maria (singular 'mare', the Latin word for 'sea'), which these areas were once believed to be. We now know that there is no water on the Moon's surface, however, and that the maria are actually plains; even so, their maritime, Latin names have stuck, including that of the largest mare, Oceanus Procellarum (Ocean of Storms), and that of perhaps the most famous, the Sea of Tranquillity (Mare Tranquillitatis). Mountain ranges (montes) ring some of the maria and craters in the lowland regions. Eighty-five per cent of the Moon's surface is made up of highlands (terrae). Impact craters are also clearly visible in these areas, which include the lunar poles. In 1959, the Soviet space probe *Luna 3* was the first to photograph the Moon's far side, revealing more, yet smaller, craters and fewer mare than on the near side.

THE LUNAR PHASES

During its 27.32-day orbit of our planet, the Moon's shape appears to change. As it moves its position relative to us, its sunlit face (and remember that the Moon merely reflects sunlight and does not generate any light of its own) disappears from our view entirely before gradually reappearing and waxing until the Moon is full, waning, disappearing and then reappearing again. The full cycle of lunar phases, or lunation, which takes 29.53 days from start to finish, begins with the new Moon. (This lunar, or synodic, month is longer than a sidereal month due to the Earth's orbit around the Sun.)

New Moon: when the Moon is positioned between the Earth and the Sun, so that the side facing us is dark.

Waxing crescent: when a visible, unlit crescent appears to be growing larger (waxing).

First quarter: when the Moon is a quarter of the way through its orbit of the Earth and we can see half of its sunlit side.

Waxing gibbous: when roughly three-quarters (gibbous) of the Moon's sunlit side is visible from the Earth.

Full Moon: when the Earth is lying between the Sun and the Moon, so that the entire sunlit side can be seen.

Waning gibbous: when about three-quarters (gibbous) of the sunlit side is visible, but this amount is gradually decreasing (waning).

Last quarter: when the Moon is three-quarters of the way through its orbit of the Earth and we can see half of its sunlit side.

Waning crescent: when the lunar crescent that is now visible is slowly dwindling in size.

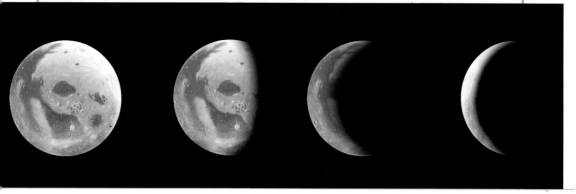

LUNAR ECLIPSES: SOME DATES FOR YOUR DIARY

A lunar eclipse is caused when the Earth passes between the Sun and the Moon, blocking the Sun's light from the Moon, which can happen up to three times a year. When the Earth's umbra falls on the Moon, a total eclipse is the result. A partial eclipse occurs when only part of the Moon is overshadowed by the Earth's umbra and the remainder is bathed in the more opaque penumbra. The eclipsed Moon may sometimes take on a brick-red hue if the Earth's atmosphere bends the Sun's rays in a certain way so that they fall on the Moon's surface.

You may be able to observe the partial eclipses that are scheduled to occur on the following dates if the Moon is above the horizon wherever it is that you happen to be:

- **17 October 2005**
visible from North America and Asia;
- **7 September 2006**
visible from eastern Africa, western Asia and India;
- **16 August 2008**
visible from eastern Europe, Africa and India;

- **21 December 2009**
visible from Europe, Africa and Asia;
- **26 June 2010**
visible from the Pacific region.

If the Moon is above your location's horizon, you may be able to witness the total lunar eclipses that will be visible from the Earth on the following dates:

- **3–4 March 2007**
visible from Europe and Africa;
- **28 August 2007**
visible from the Pacific region and eastern Australia;

- **21 February 2008**
visible from western Europe, Africa, South America and North America, but excluding the West Coast;
- **21 December 2010**
visible from North America and the Pacific region.

A crater on the Moon's surface.

Observing the Moon

The natural features of the Moon's surface can be discerned with the naked eye, and the suggestive patterns that they describe have, over the millennia, given rise to the belief in such mythical characters as the lunar rabbit of Chinese folklore and the 'man in the Moon' that Western Earthlings often glimpse.

Just looking at the Moon will enable you to recognise the dark-grey maria and the lighter-grey highlands of the near

PROFESSIONAL LUNAR OBSERVATIONS

In 1609, the English astronomer Thomas Harriot (1560–1621) is thought to have been the first person to draw up a map of the Moon based on observations made by telescope. Over the centuries that followed, increasingly sophisticated optical instruments continued to add to astronomers' lunar fact file. It was not until the Cold War era that a real breakthrough was made, however, when the Soviet Union launched its first Luna spacecraft in 1959. Luna 1 was the first spacecraft to blast through the Earth's atmosphere; Luna 2 crash-landed on the Moon in the same year; Luna 3 photographed its far side, also in 1959; and Luna 9 landed successfully in 1966 and beamed the first television images of the Moon's surface back to the Earth. It was the Apollo programme, established in 1961, that would cover the United States, rather than the Soviet Union, in glory, however.

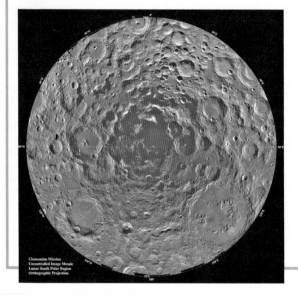

A lunar mosaic of 1,500 Clementine images of the south-polar region of the Moon. The projection is orthographic, centred on the south pole. The Schrodinger Basin (320 km, or 200 miles, in diameter) is located in the lower right of the mosaic. Amundsen-Ganswindt is the more subdued circular basin between Schrodinger and the pole. The polar regions of the Moon are of special interest because of the postulated occurrence of ice in permanently shadowed areas. The south pole is of greater interest because the area that remains in shadow is much larger than that at the north pole.

The Galileo spacecraft took these images of the Moon on 7 December 1992 on its way to explore the Jupiter system. The distinct, bright-ray crater at the bottom of the image is the Tycho impact basin. The dark areas are lava-rock-filled impact basins: Oceanus Procellarum (on the left), Mare Imbrium (centre left), Mare Serenitatis and Mare Tranquillitatis (centre), and Mare Crisium (near the right edge).

In 1969, the astronauts Neil Armstrong and Edwin (Buzz) Aldrin – the crew of Apollo 11 – not only became the first humans to walk on the Moon, but gathered rock and soil samples from the near side's surface for analysis back on the Earth. As a result, we know that the lunar regolith (soil) is made up of glassy and rocky particles and that the Moon-rock types are primarily anorthosites, basalts and breccias. Space probes have continued to investigate the Moon. The US spacecraft Clementine, while experimenting with lasers, mapped the Moon from close quarters in 1994, for example.

Stepping onto the Moon.

Human footprints on the lunar surface.

side, but swinging a pair of binoculars or a telescope in the Moon's direction will give you a much clearer view of such features as the Montes Apenninus and the bright-looking ray crater (the rays are made up of ejecta, or ejected material) called Copernicus. The best spot on which to train an optical instrument is the terminator, the imaginary line that divides night and day, or the areas of darkness and light, on the Moon, because the oblique angle at which sunlight falls on the lunar surface here sharpens the outline of, for example, mountains. And consulting a Moon map (see page 218) will help you to identify such lunar features.

Copernicus is 93 km (58 miles) wide and is located within the Mare Imbrium Basin, on the northern near-side of the Moon (10 degrees N, 20 degrees W). This image shows the crater floor, floor mounds, rim and rayed ejecta. The image was acquired on the lunar Orbiter 2 mission.

MARS
Diameter: 6,778 km (4,212 miles)
Average distance from the Sun: 227.4 million km (141.3 million miles)
Mean surface temperature: –63˚C (–81˚F)
Orbital period: 687 days
Number of moons: 2
Visible magnitude: –2.0

Mars

It was its colour that inspired the Romans to name the 'Red Planet', as it is familiarly known, after Mars, their bloodthirsty warrior god (whose equivalent in the Greek Olympic pantheon was Ares). A terrestrial, inner planet, Mars' orbit falls outside that of the Earth, which is why it is counted among the superior planets. Although its orbital period – 687 days – is nearly twice the length of the Earth's, the time that it takes to rotate once on its axis – 24.62 hours – is only slightly longer than an Earthly day. The angle at which its axis tilts from the vertical – 25.2˚ – is similarly a little greater. Like the Earth, Mars has seasons and weather, but thankfully we are spared its frenzied dust storms. The solar wind blasts past our planetary neighbour, whipping up winds with speeds of up to 300 km (186 miles) an hour and cutting through an atmosphere whose pressure is 100 times lower than that on Earth to scoop up the gritty dust on the Martian surface and hurl it around.

At the centre of this planet lies a small iron core, surrounded by a mantle of silicate rock, in turn enclosed by a thin, rocky crust. Mars' surface is covered with sand-like particles of iron oxide – literally rusted, or oxidised, iron – and this is the source of its redness rather than fiery heat. Indeed, the combination of orbiting the Sun at an average distance of 227.4 million km (141.3 million miles) and a thin atmosphere – which is rich in carbon dioxide (CO_2) – makes the Martian world a cold one. During a Martian

The Mars Global Surveyor (MGS) Mars Orbiter Camera (MOC) experiment consists of three different cameras: a narrow-angle imager that provides the black-and-white, high-resolution views (up to 1.4 m, or 4½ feet, per pixel) of Mars, and two wide-angle cameras, observing in red and blue wavelengths, from which colour views of the entire planet are assembled each day. The view of Mars shown here was assembled from MOC global images obtained on 12 May 2003. At the left/centre of this view are the four large Tharsis volcanoes: Olympus Mons, Ascraeus Mons Pavonis Mons and Arsia Mons.

The surface of Mars.

winter, around a third of the atmosphere can remain frozen above the polar icecaps. Mars is far too cold to play host to water, but this may not always have been so. Astronomers believe that the ice here has a similar composition to water ice on Earth, and that the channels visible on the planet's surface can only have been created by the erosive action of running water. They also speculate that the chaotic region of Mars, which is littered with huge rocks, may have come into being when a subterranean build-up of water broke through the surface with great violence, causing it to break up. And where there was water, there may also have been life, begging a question that has tantalised astronomers ever since William Herschel first discerned what he believed to be seas on the Red Planet: was there – is there – life on Mars?

NASA's Mars Exploration Rover Opportunity has found an iron meteorite on Mars, the first meteorite of any type ever identified on another planet. The pitted, basketball-sized object is mostly made of iron and nickel.

Mars bears the scars of the Late Heavy Bombardment (of meteoroids) in the form of impact craters and basins. One strike may have been so devastating that it created the lowlands of the northern hemisphere, for Mars' southern landscape exhibits highlands, and the difference is quite striking. While the Earth's crust comprises tectonic plates that shift in a horizontal direction, Mars' crust is thought to be immobile. This may explain the size of its volcanoes, as well as the massive extent of the plains, which were formed by lava constantly pouring from these fixed points before pooling and cooling around them. The largest of these giant volcanoes are Arsia Mons, Pavonis Mons, Ascraeus Mons and Olympus Mons. The latter is the largest volcano in the entire solar system, having an estimated height of 27 km (17 miles) and a diameter of up to 600 km (373 miles). Ascraeus Mons presides over another

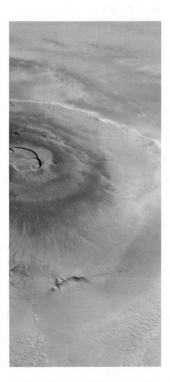

Olympus Mons is a mountain of mystery. Taller than three Mount Everests and about as wide as the entire Hawaiian-island chain, this giant volcano is nearly as flat as a pancake. That is, its flanks typically only slope 2° to 5°. The Mars Orbiter Camera (MOC) obtained this spectacular wide-angle view of Olympus Mons on Mars Global Surveyor's 263rd orbit, at around 10.40pm PDT on 25 April 1998. In the view presented here, north is to the left and east is at the top.

A Viking Orbiter 1 photomosaic of Olympus Mons' summit caldera. The caldera comprises a series of craters formed by repeated collapses after eruptions.

notable Martian feature: a rift valley 7 km (4 miles) deep, up to 600 km (373 miles) wide and around 4,500 km (2,796 miles long), which Earth-bound astronomers have named the Valles Marineris (Valley of the Mariner), after the *Mariner 9* space probe that first beamed images of it back to Earth.

Mars' moons

Mars has two natural satellites, which were discovered by the American astronomer Asaph Hall (1829–1907) in 1877, named Phobos and Deimos, after two of Ares' sons, according to Greek mythology. Thought to be asteroids captured by Mars' gravity, these dark, rocky, crater-pitted objects travel around Mars in an easterly direction. Phobos, which has a diameter of 26 km (16 miles), takes just over seven hours to orbit the Red Planet at a distance of 9,000 km (5,592 miles). Deimos, whose diameter is 16 km (nearly 10 miles), has an orbital period of a little over 30 hours and lies 23,000 km (14,292 miles) from Mars.

Observing Mars and its moons

The best time to observe the Red Planet is when it is in opposition to, or opposite, the Sun and closest to the Earth, during the alignment of Mars, the Earth and the Sun that takes place every 26 months. Mars' red glow can be discerned with the naked eye even when it is not in opposition, while a basic telescope should enable you to see its bright polar caps and patchy-looking, rocky surface, and, better still, to home in on the spectacular surface features that are located within, or near, the Tharsis Bulge, a little north of the planet's equator. Here, for instance, you may be able to identify Ascraeus Mons as a red circle, and the Valles Marineris as a darker area in its vicinity. It is very unlikely that you will be able to see Phobos and Deimos, however, because they do not reflect much light and therefore orbit Mars in near darkness.

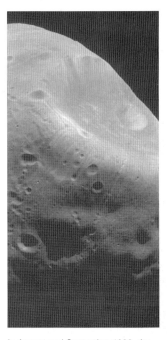

In August and September 1998, the Mars Global Surveyor Mars Orbiter Camera had four opportunities for close fly-bys of the inner moon, Phobos. This spectacular view shows the large crater Stickney towards the upper right.

PROFESSIONAL MARTIAN OBSERVATIONS

Mars has been an avidly viewed focus of astronomers' attention ever since telescopes subjected its surface to Earthlings' close scrutiny. It was the space age that enabled far more accurate and detailed observations of the Red Planet, however, especially from 1965, when the Mariner 4 spacecraft transmitted around 20 images of its bleak landscape to the United States. Since then, revelation has followed revelation, with Mariner 9 mapping the planet and opening our eyes to the Valles Marineris and Mars' titanic volcanoes between 1971 and 1972, and landers from two US Viking spacecraft descending to its surface in 1976, programmed for analysis, but producing inconclusive results. Since 1997, Mars has been rarely out of the headlines, with the US spacecraft Mars Pathfinder landing in the Ares Vallis (Valley of Ares) in 1997, where it released the Sojourner Rover robot to trundle around and send pictures of its findings, along with analyses of Martian soil and the prevailing rock types, back to an agog audience on Earth. The Mars Global Surveyor (MSG) arrived later that year, and in 1999 began its important task of surveying and mapping the planet, while NASA's Mars Exploration Rovers (MERs) Opportunity and Spirit began collecting data from the surface in 2004. Humankind's mission to Mars has continued into the 21st century, and it may not be long before we have conclusive proof either that life forms once existed on the Red Planet or that it has always been barren and uninhabited.

This eight-image mosaic was acquired during the late afternoon (note the long shadows) on Sol 2 as part of the pre-deploy 'insurance panorama' of , and shows the newly deployed rover sitting on the Martian surface. This colour image was generated from images acquired at 530,600 and 750 nm.

Scientists working with NASA's Mars Exploration Rover Spirit decided to examine this rock, dubbed 'Wishstone', based on data from the miniature thermal-emission spectrometer. Spirit used its rock-abrasion tool first to scour a patch of the rock's surface with a wire brush, and then to grind away the surface to reveal interior material.

This image from the panoramic camera on NASA's Mars Exploration Rover Opportunity features the remains of the heat shield that protected the rover from temperatures of up to 1,093°C (2,000°F) as it made its way through the Martian atmosphere. This two-frame mosaic was taken on the rover's 335th Martian day, or sol (2 January 2004).

JUPITER

Diameter: 142,664 km
(88,650 miles)
**Average distance from
the Sun:** 776.5 million km
(482.5 million miles)
**Mean cloud-top
temperature:** −110°C
(−166°F)
Orbital period: 12 years
Number of moons: 16
Visible magnitude: −2.7

Jupiter

Named for the leader of the Roman gods, whose Greek counterpart was Zeus, Jupiter is not only classified as an outer, superior planet, but as a 'gas giant', which immediately tells you that it consists primarily of gas (89.8 per cent hydrogen and 10.2 per cent helium, to be precise) and that it is gigantic. Indeed, it is the largest planet in the solar system, with a mass more than twice that of all of the other planets put together. This means that its gravitational pull has attracted many comets to it that would otherwise have hurtled towards Earth. Taking as it does just under 10 hours to make a full rotation on its axis, which is tilted 3.1° from the vertical, Jupiter also has the distinction of spinning faster than any other planet.

Unlike the terrestrial planets, gas giants have no crust, and Jupiter's outermost component is instead a dense, gaseous atmosphere 1,000 km (621 miles) deep, which envelops a layer of liquid hydrogen and helium, within which lies a layer of metallic hydrogen and helium, which in turn surrounds a solid core – perhaps a rocky one. As well as flattening its poles and causing a bulge at the equator, Jupiter's rapid rate of rotation, combined with the heat produced by its continuing contraction (its diameter was once five times larger than it is now),

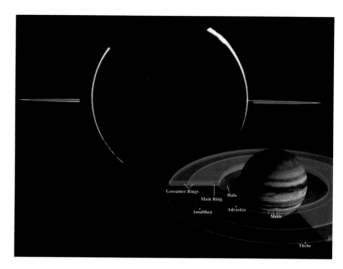

NASA's Galileo spacecraft acquired this mosaic of Jupiter's ring system when the spacecraft was in Jupiter's shadow looking back towards the Sun. Jupiter's ring system is composed of three parts: an outermost, gossamer ring, a flat, main ring and an innermost, doughnut-shaped halo. These rings are made up of dust-sized particles that have been blasted off the nearby inner satellites by small impacts. This image was taken on 9 November 1996 at a distance of 2.3 million km (1.4 million miles).

whips up colossal storms in its upper atmosphere. Indeed, one, the Great Red Spot (GRS), which is thought to measure around 15,000 by 20,000 km (9,321 by 12,428 miles), is a hurricane, or cyclone, that has been visible from Earth for hundreds of years as it circles the planet in an anti-clockwise direction a little south of the Jovian equator. Thanks to US spacecraft, we know that Jupiter is encircled by a faint ring system, which is believed to consist of dust knocked from its four inner moons by ongoing meteoroid impacts.

Jupiter's Great Red Spot.

Jupiter's moons

The first quartet of Jupiter's 16 moons to be discovered was collectively named the 'Galilean moons' in honour of Galileo, who, in 1610, was one of the first to discern them with a telescope (another was German astronomer Simon Marius). It was an important discovery, for it showed that celestial bodies did not all orbit around the Earth.

The Galilean moons' individual names are Io, Europa, Ganymede and Callisto, which, in Greek mythology, were the names of Zeus' often reluctant lovers. We now know that Jupiter has 16 moons, all of which have been named for the god's inamoratas and inamoratos, with the exception of Amalthea, who raised the infant Zeus. In order of their discovery, the 12 smaller, non-Galilean moons are: Amalthea, which was first spotted in 1892; Himalia (1904); Elara (1905); Pasiphae (1908); Sinope (1914); Lysithea and Carme (1938); Ananke (1951); Leda (1974); and Metis, Adrastea and Thebe (1979).

The US spacecraft *Voyager 1* gave Earth-bound astronomers the first images of the Galilean moons in 1979, from which they concluded that there are active volcanoes on Io. *Galileo* probed further, in 1996 revealing Europa's surface of crazed ice and Ganymede's magnetosphere, and, in 1997, providing evidence that suggests that oceans may lie beneath Europa and Callisto's icy surfaces.

Io.

An active volcanic eruption on Jupiter's moon, Io, was captured in this image taken on 22 February 2000 by NASA's Galileo spacecraft. Tvashtar Catena, a chain of giant volcanic calderas, was the location of an eruption caught in action in November 1999. A dark, 'L'-shaped lava flow to the left of centre in this, more recent, image marks the location of the November eruption. White and orange areas on the left side of the picture show newly erupted, hot lava.

The solar system's largest moon, Ganymede, is captured here alongside the planet Jupiter in a colour picture taken by NASA's Cassini spacecraft on 3 December 2000. Ganymede is larger than the planets Mercury and Pluto and Saturn's largest moon, Titan. Both Ganymede and Titan have a greater surface area than the entire Eurasian continent on Earth.

Bright scars on a darker surface testify to a long history of impacts on Jupiter's moon, Callisto, in this image of Callisto from NASA's Galileo spacecraft. The picture, taken in May 2001, is the only complete global colour image of Callisto obtained by Galileo, which has been orbiting Jupiter since December 1995. Of Jupiter's four largest moons, Callisto orbits the farthest from the planet.

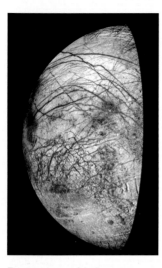

Europa, a moon of Jupiter, appears as a thick crescent in this enhanced-coloru image from NASA's Galileo spacecraft, which has been orbiting Jupiter since 1995. The view combines images taken in violet, green and near-infrared filters in 1998 and 1995. The colours have been stretched to show the subtle differences in materials that cover the icy surface of Europa. Reddish linear features are some of the cracks and ridges, thousands of kilometres long, which are caused by the tides raised by the gravitational pull of Jupiter. Mottled, reddish 'chaotic terrain' exists where the surface has been disrupted and ice blocks have moved around.

Because their orbits fall within those of Io, the first of the Galilean moons, Metis, Adrastea, Amalthea and Thebe, are called the 'inner moons'. Astronomers think that Leda, Himalia, Lysithea and Elara are fragments of an asteroid, and that Ananke, Carme, Pasiphae and Sinope are asteroids that have been captured by the planet's gravitational field.

Observing Jupiter and its moons
Jupiter appears bright enough in the night sky to spot with the naked eye, but gazing at it through a telescope will prove far more rewarding. Do this, and you should be able to see the horizontal stripes that have inspired Jupiter's informal name, the 'Banded Planet'. These bands are bright zones of rising gas and dark belts of falling gas, and it is their different molecular make-up and temperature that gives them their multi-coloured appearance. The Banded Planet's rotation is visible from the Earth, and you may also be able to locate the Great Red Spot (which nowadays looks distinctly pink) as it rages around Jupiter, typically circumnavigating a quarter of the Banded Planet in around two hours. Using a telescope (and sometimes also a pair of binoculars), you may also be able to see any, or all, of the largest of Jupiter's moons, namely Io, Europa, Ganymede and Callisto, which describe an almost perfect circle as they orbit at the level of the planet's equator.

This true-colour mosaic of Jupiter was constructed from images taken by the narrow-angle camera onboard NASA's Cassini spacecraft on 29 December 2000, during its closest approach to the giant planet at a distance of approximately 10 million km (6.2 million miles). It is the most detailed global colour portrait of Jupiter ever produced; the smallest visible features are approximately 60 km (37 miles) across. Although Cassini's camera can see more colours than humans can, Jupiter's colours in this new view look very close to how the human eye would see them.

JUPITER'S MOONS
(in order of distance from Jupiter)

	Name	Diameter	Distance from Jupiter
Inner moons	Metis	40 km (25 miles)	128,000 km (79,538 miles)
	Adrastea	20 km (12 miles)	129,000 km (80,159 miles)
	Amalthea	200 km (124 miles)	181,000 km (112,471 miles)
	Thebe	100 km (62 miles)	222,000 km (137,948 miles)
Galilean moons	Io	3,640 km (2,262 miles)	422,000 km (262,226 miles)
	Europa	3,130 km (1,945 miles)	671,000 km (416,951 miles)
	Ganymede	5,279 km (3,280 miles)	1,070,000 km (664,885 miles)
	Callisto	4,810 km (2,989 miles)	1,883,000 km (1,170,074 miles)
	Leda	10 km (6 miles)	11,094,000 km (6,893,680 miles)
	Himalia	170 km (106 miles)	11,480,000 km (7,133,536 miles)
	Lysithea	24 km (15 miles)	11,720,000 km (7,282,669 miles)
	Elara	80 km (50 miles)	11,737,000 km (7,293,233 miles)
	Ananke	20 km (12 miles)	21,200,000 km (13,173,429 miles)
	Carme	30 km (19 miles)	22,600,000 km (14,043,372 miles)
	Pasiphae	36 km (22 miles)	23,500,000 km (14,602,622 miles)
	Sinope	28 km (17 miles)	23,700,000 km (14,726,899 miles)

The Galilean satellite Io floats above the cloud tops of Jupiter in this image captured on the dawn of the new millennium, 1 January 2001, 10.00 UTC (spacecraft time), two days after Cassini's closest approach. The image is deceiving: there are 350,000 km (218,000 miles) – roughly 2.5 Jupiters – between Io and Jupiter's clouds. Io is the size of our Moon.

PROFESSIONAL JOVIAN OBSERVATIONS

The invention of the telescope heralded a significant breakthrough in humankind's understanding of the true nature of Jupiter, and 20th- and 21st-century spacecraft have paved the way to gaining even greater knowledge, starting with the US probe Pioneer 10, which travelled to the Banded Planet in 1973. Pioneer 10 established that Jupiter had a huge magnetosphere, while images transmitted to the Earth by Voyager 1 (again on a US-sponsored mission) revealed that the planet was surrounded by rings. The US spacecraft Galileo arrived at Jupiter in 1995, later dropping a probe into its atmosphere and releasing an orbiter that circled it ten times. It is thanks to Galileo that we know, for example, that the uppermost part of Jupiter's atmosphere is made up of three layers of clouds that respectively comprise ammonia ice, ammonium hydrosulphide and water and ice.

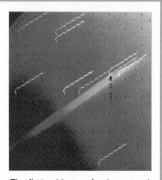

The first evidence of a ring around the planet Jupiter is seen in this photograph taken by Voyager 1 on 4 March 1979. The multiple exposure of the extremely thin, faint ring appears as a broad, light band crossing the centre of the picture.

The first discrete ammonia-ice cloud positively identified on Jupiter is shown in this image taken by NASA's Galileo spacecraft. Ammonia ice (light blue) is shown in clouds to the north-west (upper left) of the Great Red Spot (the large red spot in the middle of the image).

Jupiter and its four planet-sized moons, called the Galilean satellites, were photographed in early March 1979 by Voyager 1 for assembly into this collage. They are not to scale, but are in their relative positions.

Saturn

Saturn is the solar system's second-largest planet after
Jupiter, which is probably why the Romans named it for
Jupiter's father (Cronos in Greece), a deity of agriculture
who, according to Graeco–Roman mythology, was the
tyrannical leader of the gods until Jupiter overthrew him.
Saturn was the farthest planet visible from Earth until
William Herschel's telescope came across Uranus in 1781 –
before then, Saturn was believed to be the 'last' planet.

An outer, superior planet, Saturn takes just over 10½
hours to complete one turn on its axis, which tilts at an angle
of 26.7° from the vertical, a relatively fast rate that, when
combined with its low density (the lowest of any planet),
causes its equator to bulge dramatically. Yet when viewed

SATURN
Diameter: 120,266 km (74,732 miles)
Average distance from the Sun: 1,428 million km (887.1 million miles)
Mean cloud-top temperature: –140°C (–220°F)
Orbital period: 29.5 years
Number of known moons: 33
Visible magnitude: +0.7

Saturn's rings.

These Hubble Space Telescope images, captured from 1996 to 2000, show that Saturn's rings open up from just past edge-on to nearly fully open as it moves from autumn towards winter in its northern hemisphere.

from the Earth, the rings that circle its equator are far more striking than its equatorial bulge, so much so that Galileo, using a relatively basic telescope, was able to discern them in 1610. It was the Dutch astronomer Christiaan Huygens, however, who identified them as rings rather than a pair of moons, as Galileo had guessed. And it was the Italian astronomer Giovanni Cassini who first proposed that these rings were made up of rocks. They are now thought to be the tiny, icy remnants of fragmented comets that were captured by Saturn's gravity. Three main rings are visible from the Earth, and astronomers have labelled them 'A', 'B' and 'C' (or,

This close-up view of the lit side of Saturn's outer 'B' ring and the Cassini Division looks something like a phonograph record. There are subtle, wave-like patterns, hundreds of narrow features resembling a record's 'grooves' and a noticeable, abrupt change in overall brightness beyond the dark gap near the right. To the left of the gap is the outer 'B' ring with its sharp edge maintained by a strong gravitational resonance with the moon Mimas. To the right of the Huygens Gap are the plateau-like bands of the Cassini Division. The narrow ringlet within the gap is called the Huygens Ringlet.

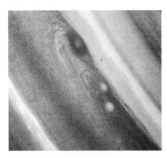

Saturn's northern hemisphere.

alternatively, the Crêpe ring): 'A' is the outermost ring; 'B', the brightest, is the central ring; and 'C', the clearest, is the inner ring.

Cassini ascertained that there is a gap – now known to be 5,000 km (3,107 miles) wide – between rings 'A' and 'B', which is now called the Cassini Division in his honour; the Encke Division, which divides the 'A' ring, was identified by Johann Encke (1791–1865), a German astronomer, in 1837. Thanks to information supplied by American space probes, such as *Pioneer 11*, which, on a fly-by, presented evidence of the 'F' ring, we now know that there is also a faint 'D' ring (which is the closest to Saturn), as well as the 'E', 'F' and 'G' rings. The outermost is the 'E' ring, which is situated about 480,000 km (298,266 miles) from Saturn. During the early 1980s, the US spacecraft *Voyager 1* and *Voyager 2* enabled astronomers to ascertain that innumerable 'ringlets' make up the main rings, and they have high hopes at the time of writing that the US spacecraft *Cassini*, which blasted off from the Earth in 1997 and reached Saturn late in 2004, will tell us even more about these, Saturn's most distinctive features.

As a gas giant, Saturn's solid core, which is probably frozen rock, is surrounded by successive layers consisting of liquid metallic hydrogen and helium, liquid hydrogen and helium and an atmosphere that also consists of hydrogen and helium. Saturn's total hydrogen–helium ratio is estimated as 96.3 per cent to 3.7 per cent. Like Jupiter's, Saturn's atmosphere is topped by three decks of clouds, the difference being that above Saturn they are lower, colder, further apart (because Saturn's gravity is less strong) and shrouded in a hazy cloud of ammonia crystals. And while storms do rage over Saturn, these typically occur at 57-year intervals – all in all, it currently seems as though Saturn's is a far calmer world than that of Jupiter.

Saturn's moons

Saturn is currently believed to have 33 moons – the greatest number of natural satellites of any planet – and astronomers have not ruled out the possibility that there may be more. Eighteen have names that originate in Greek mythology; others' names are drawn from Norse, Gallic and Inuit myths; and some, such as the two discovered in August 2004, are currently still awaiting a name. The largest moon – even bigger than Mercury – and the first to be discovered (in 1655, by Huygens) is Titan. In order of discovery, Saturn's next-biggest moons are: Iapetus (1671); Rhea (1672); Dione and Tethys (1684); Mimas and Enceladus (1789); Hyperion (1848); Phoebe (1898); Epimetheus and Janus (1966); Atlas, Prometheus, Pandora, Telesto, Calypso and Helene (1980); and Pan (1990).

In 1944, the Dutch-born US astronomer

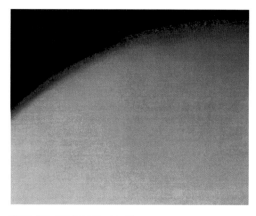

Titan, Saturn's largest moon.

These two colour composite images of Saturn's moon Iapetus, from Cassini's visual and infrared mapping spectrometer, were obtained on 31 December 2004.

This very detailed image, taken during the Cassini spacecraft's closest approach to Saturn's moon Dione on 14 December 2004 is centred on the wispy terrain of the moon. To the surprise of Cassini-imaging scientists, the wispy terrain does not consist of thick ice deposits, but rather of the bright, ice cliffs created by tectonic fractures.

Gerard Kuiper (1905–73) found that Titan had a nitrogen-rich atmosphere that blankets this moon in an orange-coloured cloud. All of Saturn's moons are believed to consist of frozen rock, and it is speculated that Titan may have oceans of ethane and methane, as well as rain. Some Saturnian natural satellites share the same orbit, namely Dione and Helene, and also Tethys, Telesto and Calypso. Prometheus and Pandora are known as 'shepherd moons' because they keep the particles within the 'F' ring together, in the manner of a shepherd shooing straying sheep back into the flock. Phoebe, the outermost moon, orbits Saturn in a different direction to the other moons.

Mimas drifts along in its orbit around Saturn against the azure backdrop of Saturn's northern latitudes in this true-colour view. The long, dark lines on the atmosphere are shadows cast by the planet's rings.

Observing Saturn and its moons

Glance up at the night sky in the right direction, and in the right conditions, and the chances are that you will glimpse the bright, yellow-tinged 'star' that is actually the planet Saturn. If you are intent on admiring its rings, however, you will require a telescope. Look through this, and the planet will assume the colour of butterscotch, thanks to its ammonia-crystal overcoat. You should also be able to discern its three main rings, unless you are unfortunate enough to train your telescope on Saturn during one of the two occasions in its nearly 30-year orbit when the rings, which are only 10 m (33 ft) thick at certain points, are edge-on to the Earth and consequently become invisible. If you are reading this, telescope primed, in September 2009, for example, be warned that you will almost certainly scour the sky in vain for the distinctive features that give the Ringed Planet its informal name. (This edge-on phenomenon has its advantages, however. When it occurred in 1966, astronomers saw Epimetheus and Janus for the first time.) Through a telescope, you should also be able to see Titan as a bright dot.

This unusual view of Saturn's moon Hyperion shows just how strangely shaped this tumbling little moon is. Hyperion is thought to be the largest irregularly shaped moon in the solar system. This moon measures 266 km (165 miles) across.

SATURN'S LARGEST MOONS
(in order of distance from Saturn)

Name	Diameter	Distance from Saturn
Pan	20 km (12 miles)	133,580 km (83,005 miles)
Atlas	34 km (21 miles)	137,640 km (85,528 miles)
Prometheus	100 km (62 miles)	139,350 km (86,590 miles)
Pandora	88 km (55 miles)	141,700 km (88,051 miles)
Epimetheus	110 km (68 miles)	151,420 km (94,091 miles)
Janus	191 km (119 miles)	151,470 km (94,122 miles)
Mimas	398 km (247 miles)	185,520 km (115,280 miles)
Enceladus	498 km (309 miles)	238,020 km (147,903 miles)
Tethys	1,060 km (659 miles)	294,660 km (183,098 miles)
Telesto	25 km (16 miles)	294,660 km (183,098 miles)
Calypso	16 km (10 miles)	294,660 km (183,098 miles)
Dione	1,120 km (696 miles)	377,400 km (234,512 miles)
Helene	32 km (20 miles)	377,400 km (234,512 miles)
Rhea	1,528 km (949 miles)	527,040 km (327,496 miles)
Titan	5,150 km (3,200 miles)	1,221,850 km (759,243 miles)
Hyperion	280 km (174 miles)	1,481,100 km (920,338 miles)
Iapetus	1,436 km (892 miles)	3,561,300 km (2,212,950 miles)
Phoebe	220 km (137 miles)	12,952,000 km (8,048,220 miles)

A view of
Enceladus.

Saturn and three of its moons.

PROFESSIONAL SATURNIAN OBSERVATIONS

Cassini has found Titan's upper atmosphere to consist of a surprising number of layers of haze, as shown in this ultraviolet image of Titan's night-side limb, colourised to look like true colour. The many fine, haze layers extend several hundred kilometres above the surface. Although this is a night-side view, with only a thin crescent receiving direct sunlight, the haze layers are bright from light scattered through the atmosphere.

Cassini captured Dione against the globe of Saturn as it approached the icy moon for its close rendezvous on 14 December 2004. This natural-colour view shows that the moon has strong variations in brightness across its surface, but a remarkable lack of colour compared to the warm hues of Saturn's atmosphere. Several oval-shaped storms are present in the planet's atmosphere, along with ripples and waves in the cloud bands.

The Cassini orbiter and Huygens probe aboard the Titan IV.

This image was returned to Earth on 14 January 2005 by the ESA's Huygens probe during its successful descent to land on Titan. This is the coloured view, following processing to add reflection-spectra data, and gives a better indication of the actual colour of the surface.

Telescopes played an important role in gathering information about Saturn, but because it is so distant from the Earth, we can really only learn more about the Ringed Planet by sending probes to it. It was only when the US spacecraft Voyager 1 flew past Saturn in 1980, for example, that the existence of the moons Atlas, Prometheus and Pandora was revealed.

At the time of writing, astronomers are counting down to Christmas Day 2004, when the NASA orbiter Cassini, which went into orbit around Saturn in June 2004 after a journey of over seven years, is scheduled to drop the European Space Agency's (ESA's) Huygens probe onto Titan's surface. The probe's mission is to investigate its alien surroundings when it lands there in mid-January 2005. The astronomical community is waiting with bated breath to see what both Cassini (which is equipped with 12 instruments of analysis and will remain at work for another four years) and Huygens will find. Stop press! Huygen's safe landing on Titan and transmission of data from its icy surface exceeded scientists' expectations. They now have a wealth of information to analyse and evaluate.

Uranus

Following William Herschel's discovery of the seventh planet in 1781, various names were considered for it – including 'Herschel's planet', 'Herschell' and 'Hershellium' – before 'Uranus' was finally settled upon. In accordance with tradition, the name is drawn from Roman mythology, Uranus (Uranos or Ouranos in Greek myth) being a sky god and the father of the Titans, including Saturn. Uranus takes just over 17 hours to spin once on its axis, which it does in a retrograde direction, and at a dramatic angle of 98° from the vertical, so that it is really orbiting the Sun on its side. This means that each of its poles is exposed to sunlight in turn, at intervals of 42 years.

Like Jupiter and Saturn, Uranus is a superior, outer planet, as well as a gas giant. Its rocky core is enclosed by a liquid mantle, comprising water, ice, ammonia and methane, which is in turn topped by an atmosphere that consists of 82.5 per cent hydrogen, 15.2 per cent helium, 2.3 per cent methane and relatively miniscule amounts of other gases.

URANUS
Diameter: 51,004 km (31,693 miles)
Average distance from the Sun: 2,864.59 million km (1,780.02 million miles)
Mean cloud-top temperature: –197°C (–323°F)
Orbital period: 84 years
Number of known moons: 24

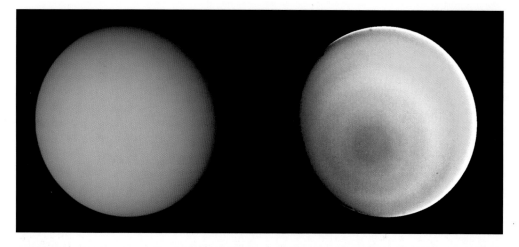

These two pictures of Uranus – one in true colour (left) and the other in false color – were compiled from images returned on 17 January 1986 by the narrow-angle camera of Voyager 2. The spacecraft was 9.1 million km (5.7 million miles) from the planet, several days from its closest approach. The picture on the left has been processed to show Uranus as human eyes would see it from the vantage point of the spacecraft. The picture is a composite of images taken through blue, green and orange filters. The darker shadings at the upper right of the disc correspond to the day–night boundary on the planet. Beyond this boundary lies the hidden, northern hemisphere of Uranus, which currently remains in total darkness as the planet rotates. The blue-green colour results from the absorption of red light by methane gas in Uranus' deep, cold and remarkably clear atmosphere.

When, during an occultation, it passed in front of a bright star in 1977, it was established that, again like Jupiter and Saturn, Uranus has rings – at least nine of them – albeit very faint ones, which circle it vertically because of the unusual position of its equatorial plane. Other features that these three planets have in common are an equatorial bulge caused by their rapid rotational speed and a host of natural satellites.

Uranus' moons

As far as we are aware, Uranus has at least 24 natural satellites, thanks to its powerful magnetic field (which is tilted at an angle of 60° to its rotational axis). Herschel sighted Titania and Oberon, the two largest, in 1781, and the remainder, in order of discovery are: Ariel and Umbriel (1851); Miranda (1948); Puck (1985); Cordelia, Ophelia, Bianca, Cressida, Desdemona, Juliet, Portia, Rosalind and Belinda (1986); Caliban and Sycorax (1997); 1986U10, Prospero, Setebos, Stephano and Trinculo (1999); and two unnamed moons discovered in 2003. Apart from 1986U10, Uranus' moons have the English playwright William Shakespeare (1564–1616) and poet and satirist Alexander Pope (1688–1744) to thank for their names, their namesakes

Above: This false-colour view of the rings of Uranus was made from images taken by Voyager 2 on 21 January 1986, from a distance of 4.17 million km (2.59 million miles). All nine known rings are visible here; the somewhat fainter, pastel lines seen between them have been contributed by the computer enhancement.

Above right: This high-resolution, colour composite of Titania was made from Voyager 2 images taken on 24 January 1986, as the spacecraft neared its closest approach to Uranus. The spacecraft was about 500,000 km (300,000 miles) away; the picture shows details about 9 km (6 miles) in size. In addition to many scars due to impacts, Titania displays evidence of other geologic activity at some point in its history. The large, trench-like feature near the terminator (the day–night boundary) at middle right suggests at least one episode of tectonic activity.

Below right: This Voyager 2 picture of Oberon is the best that the spacecraft acquired of Uranus' second largest moon. The picture was taken on 24 January 1986, from a distance of 660,000 km (410,000 miles). Clearly visible on Oberon's icy surface are several large impact craters, surrounded by bright rays similar to those seen on Jupiter's moon, Callisto.

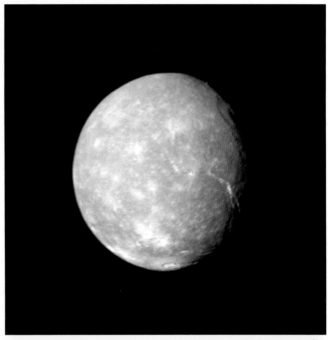

URANUS' LARGEST MOONS
(in order of distance from Uranus)

Name	Diameter	Distance from Uranus
Cordelia	26 km (16 miles)	50,000 km (31,069 miles)
Ophelia	32 km (20 miles)	54,000 km (33,555 miles)
Bianca	44 km (27 miles)	59,000 km (36,662 miles)
Cressida	66 km (41 miles)	62,000 km (38,526 miles)
Desdemona	58 km (36 miles)	63,000 km (39,147 miles)
Juliet	84 km (52 miles)	64,000 km (39,769 miles)
Portia	110 km (68 miles)	66,000 km (41,012 miles)
Rosalind	54 km (34 miles)	70,000 km (43,497 miles)
Belinda	68 km (42 miles)	75,000 km (46,604 miles)
1986U10	40 km (25 miles)	75,000 km (46,604 miles)
Puck	154 km (96 miles)	86,000 km (5,3439 miles)
Miranda	472 km (293 miles)	130,000 km (80,780 miles)
Ariel	1,158 km (720 miles)	191,000 km (118,685 miles)
Umbriel	1,170 km (727 miles)	266,000 km (165,289 miles)
Titania	1,578 km (981 miles)	436,000 km (270,925 miles)
Oberon	1,523 km (946 miles)	583,000 km (362,269 miles)
Caliban	80 km (50 miles)	7,200,000 km (4,473,995 miles)
Sycorax	60 km (37 miles)	12,200,000 km (7,580,936 miles)

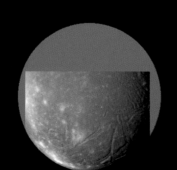
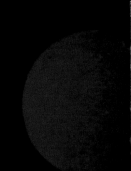

all being characters in these writers' works.

Titania, Oberon, Umbriel, Ariel and Miranda are significantly larger than the others, which is why, being more visible, they were discovered earlier. All five are cratered to a greater or lesser extent, and notable features include the bright ring (probably a crater), called a 'fluorescent Cheerio', at Umbriel's north pole and the 'V'-shaped groove, called the Inverness Corona, or chevron, that is visible on Miranda.

In 1986, *Voyager 2* ascertained that Cordelia and Ophelia were shepherd moons that constrained the limit of Uranus' outermost ring, the Epsilon ring. And Caliban and Sycorax orbit Uranus in the opposite direction to the other moons.

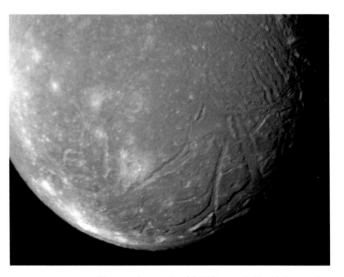

The complex terrain of Ariel is revealed in this image, the best Voyager 2 colour picture of the Uranian moon. The photos used to construct this composite were taken on 24 January 1986, from a distance of 170,000 km (105,000 miles). Most of the visible surface consists of relatively intensely cratered terrain transected by fault scarps and fault-bounded valleys (graben).

A montage of Uranus' five largest satellites. From top to bottom, in order of decreasing distance from Uranus, are Oberon, Titania, Umbriel, Ariel and Miranda. Images are presented to show their correct relative sizes and brightness. Note that coverage is incomplete for Miranda and Ariel; grey circles depict missing areas.

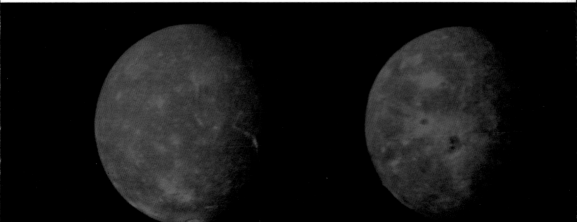

Processing this image has brought out Uranus' atmosphere.

Observing Uranus and its moons
Before the invention of telescopes, no one knew that Uranus existed, which is why you will need a pair of binoculars or a telescope if you want to observe this planet for yourself; the stronger the magnification, the more you will see. That said, you will probably only perceive a blue-green disc, the colour being provided by the Uranian clouds, which are tinted blue-green by the icy crystals of methane that they contain. Uranus' rings are so dark that you are unlikely to discern them, and the same goes for Umbriel, although you may at least be rewarded with a sight of Ariel, the brightest Uranian moon, and perhaps also of Titania, Oberon and Miranda.

PROFESSIONAL URANIAN OBSERVATIONS

Uranus is so far away from Earth that only space observation will enable astronomers to add significantly to their current stock of Uranian knowledge. Indeed, the process has already begun, for on a fly-by between 1985 and 1986, the US spacecraft Voyager 2 provided evidence that Uranus had ten more moons than had previously been identified.

An artist's concept of Voyager 2.

Neptune

Even though its level is estimated at only 1 per cent, there is so much methane in the eighth planet's atmosphere that it appears blue when viewed from space. This is one of the reasons why astronomers eventually settled on 'Neptune' as the name of the planet that Johann Galle discovered (with the help of Urbain Le Verrier, who had suggested that an unknown planet was affecting Uranus' orbit) in 1846, Neptune being the Roman god of the deep, blue seas and oceans (whose Greek counterpart was Poseidon).

This superior, outer planet is the smallest of the gas giants. It rotates on its axis at an angle of 28.3° from the vertical, taking just over 16 hours to complete one full turn. As well as methane and traces of other gases, its atmosphere consists of 80 per cent hydrogen and 19 per cent helium;

NEPTUNE
Diameter: 49,381 km (30,685 miles)
Average distance from the Sun: 4,570 million km (2,839.6 million miles)
Mean cloud-top temperature: −200°C (−328°F)
Orbital period: 165 years
Number of known moons: 13

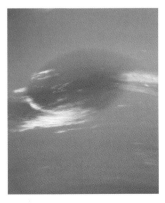

Neptune's Great Dark Spot photographed in high resolution.

Above: This image of Neptune was taken by Voyager 2's wide-angle camera when the spacecraft was 590,000 km (370,000 miles) away. The image has been processed to obtain a true-colour balance.

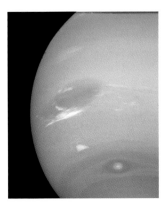

This photograph of Neptune was reconstructed from two images taken by Voyager 2's narrow-angle camera, through the green and clear filters. The image shows three of the features that Voyager 2 has consistently been photographing. At the north (top) is the Great Dark Spot, accompanied by bright, white clouds undergoing rapid changes in appearance. To the south of the Great Dark Spot is the bright feature that scientists nicknamed 'the Scooter'. Still farther south is the feature called 'Dark Spot 2', which has a bright core. Each feature moved eastward at a different velocity, so it was only occasionally that they appeared close to each other, such as when this picture was taken.

underneath this atmosphere lies an ice-and-liquid mantle comprising water, methane and ammonia that cocoons a rocky core. Neptune has quite turbulent weather – we know this because in 1989 the US spacecraft *Voyager 2* photographed bright clouds casting shadows on the main deck of clouds and measured wind speeds of more than 580 km (360 miles) an hour. *Voyager 2* also beamed back an image of a dark, Earth-sized, oval storm cloud, which astronomers called the Great Dark Spot, that visibly circumnavigated the planet over a period of 16 days. This had disappeared by the time the Hubble Space Telescope photographed Neptune in 1994, however. Another *Voyager-2*-witnessed feature to excite Earth-bound astronomers' interest was a shape-shifting area of bright cloud – another storm cloud – that zipped around the planet, thereby earning itself the designation 'the Scooter'. Two further reasons why astronomers bless *Voyager 2* are, first, that it confirmed that Neptune was encircled by a series of faint rings (whose existence had been suspected since at least 1984) and, secondly, that it gave Earthlings a clearer view of the planet's moons – indeed, six of them were revealed for the very first time.

Neptune's moons

Neptune has 13 known moons, which have been named for some of the minor sea deities, nymphs and creatures that populate Greek mythology. Triton – the largest moon – was discovered by the English astronomer William Lassell

This colour photo of Neptune's large satellite, Triton, was obtained on 24 August 1989 at a range of 530,000 km (330,000 miles). The resolution is about 10 km (6.2 miles), sufficient to begin to show topographic detail.

A composite view showing Neptune on Triton's horizon. Neptune's south pole is to the left; clearly visible in the planet's southern hemisphere is the Great Dark Spot, a large anti-cyclonic storm system located about 20 degrees south. The foreground is a computer-generated view of Triton's maria as they would appear from a point approximately 45 km (28 miles) above the surface. The terraces visible in this image indicate multiple episodes of 'cryo-volcanic' flooding.

(1799–1880) in 1846, while Gerard Kuiper was the first to discern Nereid in 1949. Thereafter, it was not until 1989, when Voyager 2 arrived at Neptune, that astronomers learned of the existence of Naiad, Thalassa, Despina, Galatea, Larissa and Proteus. Five more moons have come to light since then, and there are probably still more that haven't yet been noticed.

Of Neptune's natural satellites, Triton is one of the most interesting, for it orbits its parent planet in the opposite direction to its fellows. Triton is larger than Pluto, and, at −235°C (−391°F), has the coldest surface of any celestial body whose temperature has yet been measured in the solar system. Nereid is also unusual in having the most extreme and eccentric elliptical orbit of any celestial object: the closest it comes to Neptune is 1,353,600 km (8,411,111 miles), while its furthest point is 9,623,700 km (598,005 miles) away. It is speculated that both of these moons may be asteroids that were captured within Neptune's gravitational field.

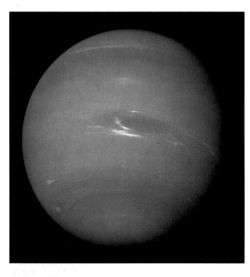

Observing Neptune and its moons

Even when your natural vision is boosted with a pair of high-magnification binoculars or a telescope, you will sadly still only see Neptune as a faint, blue-green dot in the night sky, while its moons will remain invisible to you.

This picture of Neptune was produced from the last whole-planet images taken through the green and orange filters on the Voyager 2 narrow-angle camera. The images were taken at a range of 7 million km (4.4 million miles) from the planet. The picture shows the Great Dark Spot and its companion bright smudge; on the west limb, the fast-moving bright feature called 'the Scooter' and the little dark spot are visible. These clouds were seen to persist for as long as Voyager's cameras could resolve them. North of these, a bright cloud band similar to the south polar streak may be seen.

NEPTUNE'S LARGEST MOONS
(in order of distance from Neptune)

Name	Diameter	Distance from Neptune
Naiad	58 km (36 miles)	48,230 km (29,970 miles)
Thalassa	80 km (50 miles)	50,070 km (31,113 miles)
Despina	148 km (92 miles)	52,530 km (32,642 miles)
Galatea	158 km (98 miles)	61,950 km (38,495 miles)
Larissa	104 x 89 km (65 x 55 miles)	73,550 km (45,703 miles)
Proteus	218 x 208 x 201 km (135 x 129 x 125 miles)	117,650 km (73,106 miles)
Triton	2,704 km (1,680 miles)	354,760 km (220,444 miles)
Nereid	340 km (211 miles)	5,513,400 km (3,425,962 miles)

PROFESSIONAL NEPTUNIAN OBSERVATIONS

Sending an observatory into space is the next step in probing Neptune and its moons for further information. This is bound to happen in the not too distant future.

Pluto
Hubble Space Telescope · Faint Object Camera

PRC96-09a · ST ScI OPO · March 7, 1996 · A. Stern (SwRI), M. Buie (Lowell Obs.), NASA, ESA

Pluto

Although Clyde Tombaugh was the first to pinpoint Pluto's position, in 1930, Percival Lowell had already predicted the existence of 'Planet X', as he provisionally called it. This distant planet was named for the Roman god of the underworld (Dis, or Hades, in Greece), whose dark and gloomy realm was inhabited by the souls of the dead. Pluto is such a tiny planet (smaller than our Moon), and its orbit is so eccentrically irregular in comparison to the others, that many astronomers question its planetary status, considering it more likely to be a Kuiper Belt object (see page 126). The International Astronomical Union has designated it a planet (which is usually defined as the largest object in its region of space), however, and a planet it will therefore remain unless

PLUTO
Diameter: 2,314 km (1,438 miles)
Average distance from the Sun: 5,901 million km (3,666.66 million miles)
Mean cloud-top temperature: –223˚C (–369˚F)
Orbital period: 248 years
Number of known moons: 1

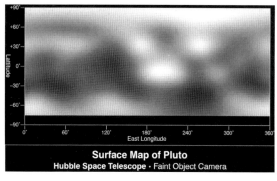

Surface Map of Pluto
Hubble Space Telescope · Faint Object Camera

This is the first image-based surface map of the solar system's most remote planet, Pluto. This map was assembled by computer image-processing software from four separate images of Pluto's disc taken with the European Space Agency's Faint Object Camera (FOC) aboard NASA's Hubble Space Telescope. Hubble imaged nearly the entire surface as Pluto rotated on its axis in late June and early July 1994.

This is the clearest view yet of the distant planet Pluto and its moon, Charon, as revealed by NASA's Hubble Space Telescope. The image was taken by the European Space Agency's Faint Object Camera on 21 February 1994 when the planet was 4.4 billion km (2.6 billion miles) from Earth, or nearly 30 times the separation between Earth and the Sun.

incontrovertible evidence is found to prove that it is another type of celestial body.

So, for the moment at least, Pluto is classed as a superior, outer planet that tilts on its axis at an angle of 122° from the vertical as it describes a full circle in nearly 6½ days. During the roughly 248 years that it takes to orbit the Sun, Pluto's position varies wildly, so that it is sometimes closer to the Sun than Neptune, but is the most distant planet from the Sun at other times. A mantle of water ice is believed to surround its rocky core, and its outermost layer is probably made up of frozen methane and nitrogen. This highly reflective layer melts a little when it is closest to the Sun, creating a thin atmosphere that then freezes again, condenses and drops back to the surface as the planet moves away from the Sun's heat.

Pluto's moon

In 1978, the American astronomer James Christy was the first Earthling to spot Pluto's moon. His discovery was named Charon, for the boatman who ferried the souls of the dead across the river Styx in Hades, according Greek mythology. Thought to be a chip off the old block, or a piece of ice that flew off Pluto and went into orbit around it when an object collided with the planet, Charon measures 1,172 km (789 miles) in diameter – remarkably similar in size to Pluto, which is why the two

are sometimes called a double-planet system. The same sides of both the planet and moon always face one another as Charon circles Pluto at an average distance of 19,640 km (12,204 miles).

Observing Pluto and its moon
Neither Pluto nor Charon is visible to amateur astronomers, even with a powerful pair of binoculars or a telescope.

PROFESSIONAL PLUTONIAN OBSERVATIONS

Pluto and Charon are so distant from us, and so similar in size, that professional astronomers using large, Earth-based telescopes can only perceive them as faint dots, and also have difficulty telling them apart. The Hubble Space Telescope (HST) has sighted bright markings on Pluto's surface, which may be icy craters, and has ascertained that the surface itself is dark red, and that Charon is redder than Pluto.

Plans are afoot to observe Pluto and Charon from an even closer position than the HST can manage. The spacecraft New Horizons will take around a decade to reach Pluto after it blasts off from the USA in 2006, but NASA scientists are hoping that the wait will be worth it and that it will send back a raft of information on Pluto and its natural satellite.

Sedna

The discovery, in 2003, by a NASA team led by Mike Brown, based at the Palomar Observatory in California, of a red, Pluto-like object, with an estimated diameter of 1,770 km (1,100 miles), that is orbiting the Sun beyond Pluto, has raised all sorts of questions about the nine-planet model of our solar system. Is this celestial object – which was named Sedna, after the Inuit sea goddess – a planet? If so, it is the tenth planet in our solar system. And if it is not a planet, then neither, in all likelihood, is Pluto. Maybe it is a Kuiper Belt object (see page 126), but if it is, it is significantly larger than the largest-known to date, namely Quaoar, whose diameter is 1,287 km (800 miles). For the record, Sedna, which may have a moon, follows an elliptical orbit around the Sun, at a distance that varies between 13 and 209 billion km (8 and 130 billion miles), taking 10,500 years to complete its solar circuit.

These four panels show the location of the newly discovered planet-like object dubbed 'Sedna' which lies in the farthest reaches of our solar system. Each panel, moving anti-clockwise from the upper left, successively zooms out to place Sedna in context. The first panel shows the orbits of the inner planets, including Earth, and the Asteroid Belt that lies between Mars and Jupiter. In the second panel, Sedna is shown well outside the orbits of the outer planets and the more distant Kuiper Belt objects. Sedna's full orbit is illustrated in the third panel, along with the object's current location. Sedna is nearing its closest approach to the Sun; its 10,000 year orbit typically takes it to far greater distances. The final panel zooms out much farther, showing that even this large, elliptical orbit falls inside what was previously thought to be the inner edge of the Oort Cloud. The Oort Cloud is a spherical distribution of cold, icy bodies lying at the limits of the Sun's gravitational pull. Sedna's presence suggests that this Oort Cloud is much closer than scientists had once believed.

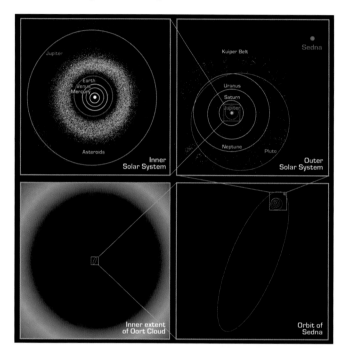

In this artist's visualisation, the newly discovered planet-like object, dubbed 'Sedna', is shown where it resides at the outer edges of the known solar system. The object is so far away that the Sun appears as an extremely bright star instead of a large, warm disc observed from Earth. All that is known about Sedna's appearance is that it has a reddish hue, and that it is almost as red and reflective as Mars. In the distance is a hypothetical small moon, which scientists believe may be orbiting this distant body.

THE MINOR MEMBERS OF THE SOLAR SYSTEM

As well as planets and their moons, the solar system is home to innumerable 'minor members', rocky objects that often also comprise dust, ice and snow. These include meteoroids, which sometimes fall to Earth; asteroids, which inhabit the Asteroid Belt between the orbits of Mars and Jupiter, that is, within the inner solar system; the comet-like objects that can be found within the more distant Kuiper Belt; and the comets with which the Oort Cloud is teeming.

All minor members orbit the Sun. Some may cross the Asteroid and Kuiper belts and hurtle towards the Sun. In so doing, they occasionally pass perilously close to the Earth – and it is thought to have been an asteroid's collision with the Earth that was ultimately responsible for wiping out the dinosaurs. Although most meteoroids – the bodies that most commonly enter the Earth's atmosphere – break up between 10 and 30 km (6 and 19 miles) above the Earth's surface, about 3,000 a year are nevertheless thought to score a direct hit on our planet.

Meteoroids and their ilk

You may be confused about the difference between a
meteoroid, a meteor and a meteorite. If so:

● a meteoroid is a small fragment of either an asteroid that
was ripped from its parent when another asteroid cannoned
into it or a short-period comet that shed it when travelling
close to the Sun; meteoroids can range in size and weight
from anything between that of a dust mote to 1,000 kg
(2,205 lb);

● a meteor is the blazing trail of light emitted by a tiny
meteoroid that is speeding through the Earth's atmosphere at
a rate of between 10 and 75 km (6 and 47 miles) an hour; as
friction causes its rocky, dusty body to heat up, evaporate
and glow, it becomes ever hotter and brighter; its trail may
measure 7–20 km (4–12 miles) in length and 1 m (3¼ feet) in
width; and

● a meteorite is the remnant of a meteoroid, and ultimately
of an asteroid or comet, that has crashed into a planet's

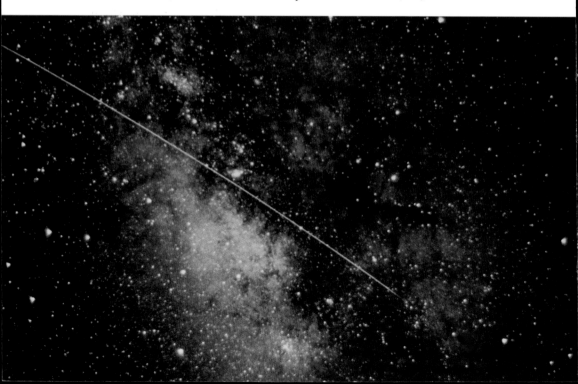

A meteor streak running through the
Milky Way.

A crater created by a fallen meteorite in Arizona.

surface, often resulting in an impact crater or basin; meteorites are classed as either 'stones' (when they are made of stone), 'irons' (when they are composed of iron and nickel) or 'stony irons' (when they comprise both stone and iron); some of the meteorites that have landed on the Earth are thought to have been chipped from the surface of the Moon and Mars by asteroid or meteoroid impacts because they share these celestial bodies' chemical composition.

Observing meteors and meteor showers
Knowing that meteors are also commonly termed 'shooting stars' and 'falling stars' should tell you what they look like when we Earthlings catch sight of them in the night sky. With magnitudes of between +3.75 and +0.75, these bright streaks are usually easily discernible with the naked eye at heights of between 80 and 120 km (50 and 75 miles) above the Earth, when they are visible for a second or so. That said, really large meteoroids generate longer-lasting meteors, some of which, at around –5 in magnitude, are so bright that they merit the name 'fireball'. These can be seen lower in the sky than the meteors generated by smaller meteoroids, and if they break up – generating a sonic boom – they are termed 'bolides'.

It is possible to observe either random meteors or meteor showers, which occur when a short-period comet passes by, trailing a stream of meteors in its wake. When the Earth crosses its orbit and passes through the 'swarm', we are privileged to marvel at the sight of a hail of meteors illuminated against the dark sky. Meteor showers typically last for several nights, increasing, and then decreasing, in intensity.

Rather like the water that spurts from a bathroom's showerhead, all meteor-shower members emanate from a single point, called the radiant. And the names of such meteor showers as the Leonids are derived from the constellation in which the radiant lies, which, in the case of the Leonids, is Leo. The comet responsible for this particular annual meteor shower is Comet 55P/Tempel-Tuttle. (Note that meteor showers can also be named after an associated comet.)

This image taken by NASA's Spitzer Space Telescope shows the comet Encke riding along its pebbly trail of debris (the long, diagonal line) between the orbits of Mars and Jupiter. This material actually encircles the solar system, following the path of Encke's orbit. Twin jets of material can also be seen shooting away from the comet in the short, fan-shaped emission that is spreading horizontally from the comet.

Encke, which orbits the Sun every 3.3 years, is well travelled. Having exhausted its supply of fine particles, it now leaves a long trail of larger, more gravel-like debris, about 1 mm (0.04 inch) in size or greater. Every October, Earth passes through Encke's wake, when we can see the Taurid meteor shower.

SOME ANNUAL METEOR SHOWERS

Name	Radiant in	Approximate dates	Where best observed
Quadrantids	Boötes	1–6 January	northern hemisphere
April Lyrids	Lyra	19–25 April	both hemispheres
Eta Aquarids	Aquarius	1–8 May	both hemispheres
Delta Aquarids	Aquarius	15 July–15 August	both hemispheres
Perseids	Perseus	25 July–18 August	northern hemisphere
Orionids	Orion	16–27 October	both hemispheres
Leonids	Leo	15–20 November	both hemispheres
Geminids	Gemini	7–15 December	both hemispheres

Asteroids

Asteroids, which are sometimes called minor planets or planetoids, are space rocks that orbit the Sun, mainly within the Asteroid Belt, which lies between the orbits of Mars and Jupiter at a distance of between 254 and 598 million km (158 and 372 million miles) from the Sun.

From analysing meteorites that have fallen to Earth, astronomers have ascertained that asteroids appear to be divided into three types:

- M-type asteroids, which are mainly made up of metal (iron and nickel);
- C-type asteroids, which contain carbon; and
- S-type asteroids, which comprise a mixture of metal and rock.

The diameters of asteroids can measure between a few metres (or feet) and 932 km (579 miles), which is the estimated diameter of Ceres, the largest asteroid identified so far, and also the first to be discovered when it was spotted in 1801 by the Italian astronomer Giuseppe Piazzi (1746–1826). Ceres, like all asteroids, spins on its axis, in its case, approximately once every nine hours. Smaller asteroids are irregular in shape, while larger ones tend to be spherical.

NEAR Shoemaker took these images of Eros on 16 October 2000, while orbiting 54 km (34 miles) above the asteroid. They are shown in false colour, constructed from images taken in green light and two different wavelengths of infrared light. Surface materials that have been darkened and reddened by the solar wind and micrometeorite impacts appear pale brown, whereas fresher materials exposed from the subsurface on steep slopes appear bright white or blue. Compared with Gaspra and Ida, similar asteroids imaged in colour from the Galileo spacecraft, Eros, exhibit large brightness variations, but only subtle colour variations.

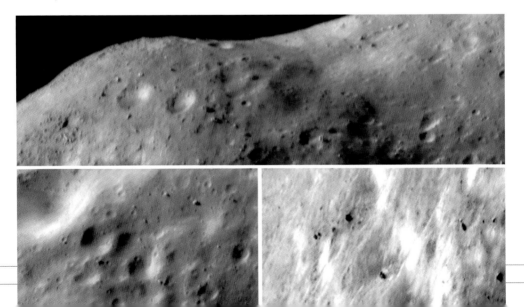

A digital representation of an asteroid field.

The Asteroid Belt is home to billions of these space rocks, which follow individual orbits and typically complete a circuit of the Sun (moving in the same direction as the planets) in anything between three and six years. But the Asteroid Belt once held around 650 larger protoplanets, which failed to conglomerate to form a planet and were instead jumbled together by the force of Jupiter's gravity, repeatedly colliding with one another until they had evolved into the asteroids that we see today. Collisions still occur, resulting variously in cratering, destruction or the creation of a 'family' of asteroids, such as that named Flora.

Jupiter continues to exert an influence on certain asteroids. The Trojan group shares its orbit, and the Asteroid Belt's so-called 'Kirkwood gaps' are kept clear of asteroids by the planet's gravitational force, which, along with that of Mars, compels some asteroids to orbit the Sun nearer to the Earth than the realm of the Asteroid Belt. These near-Earth asteroids (NEAs) are the components of the Amor group, which crosses Mars' orbit; members of the Aten group, which tends to remain within the Earth's orbit; and the Apollo asteroids, which cross the Earth's orbit.

This set of colour images of asteroid 243 Ida was taken by the imaging system on the Galileo spacecraft as it approached and raced past the asteroid on 28 August 1993.

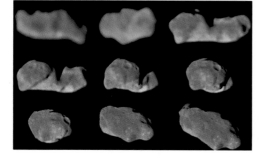

Another group – the Centaurs, of which the largest is Chiron – orbits the Sun between Jupiter and Neptune's orbit, but astronomers suspect that these icy objects may actually be the nuclei of comets rather than asteroids.

Finally, if you're wondering what asteroids are named after, the answer is that their discoverers may call them anything that they choose, subject to the International Astronomical Union's approval. Asteroids are also catalogued by year (of discovery), when they are distinguished by individually designated letters.

Observing asteroids

Of the 'Big Four' asteroids, Ceres, Pallas, Juno and Vesta, Vesta is just about visible from Earth with the naked eye. However, even a high-magnification telescope will only enable you to see these large asteroids as specks of light resembling stars (and 'asteroid' is derived from the Greek for 'star-like').

PROFESSIONAL OBSERVATIONS OF ASTEROIDS

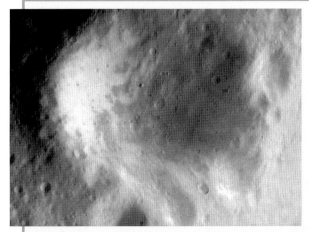

This view of the largest crater on Eros – a mosaic of NEAR Shoemaker images taken 10 September 2000, from an altitude of 100 km (62 miles) – offers a new perspective on the feature known as Psyche. The images were taken as the spacecraft flew directly over the 5.3 km- (3.3-mile-) wide crater and its smaller sister craters, which align its rim and create a paw-like appearance.

Optical telescopes gave professional astronomers their first glimpse of asteroids, but it is both the examination of meteorites and the scrutiny of space probes that have widened our knowledge of them. NASA's Near-Earth Asteroid Rendezvous (NEAR) Shoemaker, which was launched in 1997, flew past Mathilde and landed on Eros in 2001, for example, photographing and analysing these asteroids.

Finally, a vital reason why astronomers scan the sky looking for NEAs is that there is a definite danger that one of these asteroids may slip from the Asteroid Belt, if it hasn't already, and, enticed by the Sun's gravitational field, smash into Earth one day!

Kuiper Belt objects

The Kuiper Belt extends between Neptune's orbit and the Oort Cloud. It is sometimes also known as the Edgeworth–Kuiper Belt, in honour of the work carried out by the British astronomers Kenneth E. Edgeworth (1880–1972) and Gerard Kuiper that led to confirmation of its existence in 1992. This region contains objects that orbit the Sun, but because they are so distant, we know little about them. They appear to resemble comets (indeed, some may be short-period comets), and the largest Kuiper Belt object (KBO) yet spotted is Quaoar, which has a diameter of 1,287 km (800 miles). (See also Pluto, pages 115-7, and Sedna, page 118.)

Observing Kuiper Belt objects

Telescope or no telescope, you won't be able to identify a Kuiper Belt object from the Earth. Even professional astronomers have difficulty discerning them, and rely on charge-coupled devices (CCDs) to capture the dim light that they radiate.

This artist's rendition shows the newly discovered planet-like object, Sedna in relation to other bodies in the solar system, including Earth and its Moon; Pluto; and Quaoar, a KBO that was until now the largest-known object beyond Pluto. The diameter of Sedna is slightly smaller than Pluto's but probably somewhat larger than Quaoar's.

Sedna
800-1100 miles in diameter

Quaoar
(800 miles)

Pluto
(1400 miles)

Moon
(2100 miles)

Earth
(8000 miles)

Comets

Nicknamed 'dirty snowballs', comets are snowy, icy, rocky objects that are mainly confined to a spherical region named the Oort Cloud, for the Dutch astrophysicist Jan Oort (1900–92) who first proposed its existence. It is thought that 10 trillion comets orbit the Sun (following individual orbits) within the Oort Cloud, which lies beyond Pluto's orbit and is estimated to extend for 7.6 million km (4.7 million miles). The Oort Cloud's outer edge falls short of the nearest star by roughly four-fifths of the distance between them. It takes each comet around 10 million years to complete a lap around the Sun.

Comets are irregularly shaped, which is why a comet's leading edge is said to be a 'head', at the centre of which lies a nucleus consisting of frozen ice and gas. And when, as occasionally happens, a comet strays from the Oort Cloud and passes through the Kuiper Belt into the inner solar system, the ever-increasing heat from the Sun causes the material within the nucleus to vaporise and expand, resulting in a large, glowing head, or coma, measuring up to 100,000 km (62,139 miles) in diameter, and – thanks to the action of both radiation and the solar wind – two tails streaming out behind it. One tail consists of gas and looks like a straight, narrow, blue streak; the other comprises dust particles and appears more curved, wider and yellowish-white in colour. The nearer the comet is to the Sun, the longer the tails; their average length generally being 100 million km (62 million miles).

The comets that leave the Oort Cloud and embark on a voyage towards the Sun on a different, extremely elliptical, orbit are called periodic comets because they become visible to Earth-bound astronomers at regular intervals. The intervals between the reappearance of long-period comets are often thousands of years (in Comet P/Hale–Bopp's case, around 6,580 years). As their name suggests, short-period comets return to the Sun, and consequently to skies above

Halley's Comet.

Comet P/Hale–Bopp.

the Earth, sooner, and usually in less than two centuries. Halley's Comet, for instance, returns every 76 years.

Comets do not last forever. It is estimated that a comet in the Oort Cloud will orbit the Sun around 1,000 times before its dwindling loss of mass results in it burning up. A comet that has entered the inner solar system will probably only manage 100 or so orbits before dying. Alternatively, it may suffer a similar fate to the one that befell Comet P/Shoemaker–Levy 9 (named for Carolyn and Gene Shoemaker and David Levy, who discovered it in 1993), which was captured and torn apart by Jupiter's gravity, causing 21 fragments to collide with Jupiter in 1994.

Apart from Halley's Comet (which was named for Edmond Halley, who predicted its existence by computing its orbit), comets are both named after the astronomers who discovered them and given a catalogue number made up of identifying letters and their year of discovery. Thus Hale–Bopp (as it is informally known), which was discovered in 1995 by the US astronomers Alan Hale, in New Mexico, and Thomas Bopp, in Arizona, is also designated C/1995 O1. In formal terminology, periodic comets are denoted by the insertion of 'P/' before their names.

Observing comets

Long- and short-period comets, i.e., those that are approaching the Sun, are often visible with the naked eye when they pass close to the Earth. You will need a pair of binoculars or a telescope to track the progress of others, however, when they will appear as bright, blurry objects against the dark backdrop of the night sky.

DATES FOR YOUR DIARY

If you can, train your eye, binoculars or telescope on the following constellations, on the following dates, and you may just see a periodic comet on its passage past Earth.

Constellation	Date	Comet	Magnitude
Hercules	May 2006	Schwassmann–Wachmann 3	+1.5
Aries	June 2006	Honda–Mrkos–Pajdusakova	+10
Pisces	November 2006	Faye	+10
Pisces	January 2008	Tuttle	+5
Aquarius	July 2009	Kopff	+9
Virgo	March 2010	Wild 2	+8.5
Cetus	July 2010	Tempel 2	+8
Gemini	October 2010	Hartley	+5

PROFESSIONAL OBSERVATIONS OF COMETS

Comet Wild 2 – an analglyphic image.

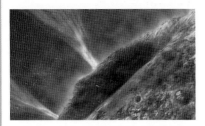

Above: This is an artist's concept depicting a view of Comet Wild 2 as seen from NASA's Stardust spacecraft during its fly-by of the comet on 2 January 2004.

Ancient star-gazers recorded the appearance of comets in our skies thousands of years ago (see page 11), while the advent of telescopes enabled astronomers to observe comets in unprecedented detail. The maturing space age has dramatically furthered our cometary knowledge. For example, there was huge excitement among astronomers in 1986, when the European Space Agency's Giotto spacecraft beamed back images of Halley's Comet's snowy nucleus; it was the first time that one had been seen at close quarters. Giotto also measured it, finding it to be 16 km (10 miles) long. In 2004, the space probe Stardust arrived within 236 km (147 miles) of Comet Wild 2 and gathered sample particles in its aerogel collector for NASA scientists to analyse when it returns to Earth. This is scheduled for January 2006, the probe already having delivered surprises in the form of images of craters, cliffs and jets of molten material on the comet's surface. And in 2005, a drop-probe impactor aboard the NASA spacecraft Deep Impact will be directed at Comet Tempel 1, whereupon the mother ship will gather information relating to the impact and resulting crater in order to ascertain the thickness of the comet's crust.

Deep Impact awaits launch from Cape Canaveral Air Force Station on 12 January 2005. The spacecraft will travel to Comet Tempel 1 and will release an impactor, creating a crater on the surface of the comet. Scientists believe that the exposed materials may give clues to the formation of our solar system.

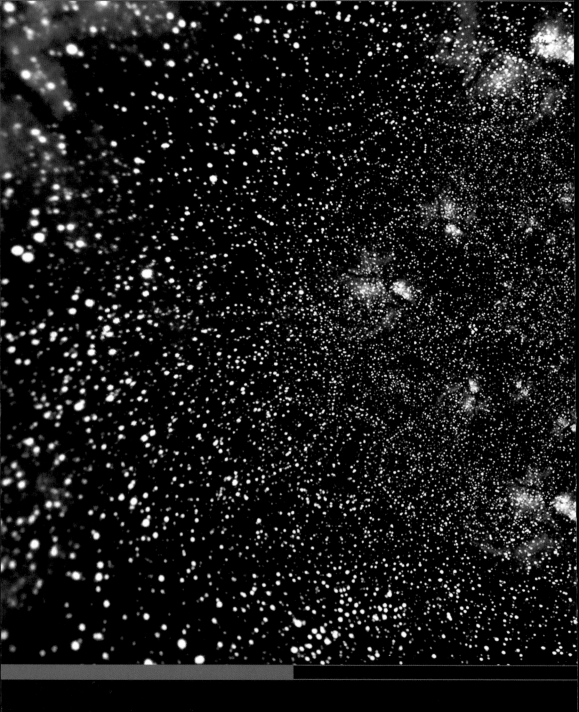

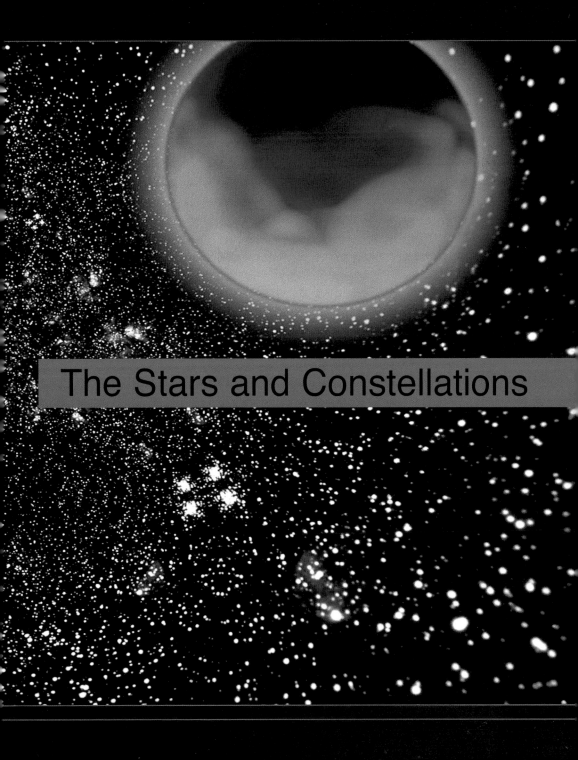

The Stars and Constellations

The Sun is both literally and metaphorically the 'star' of our solar system, yet it is nothing special in the universal scheme of things, being one of millions of stars, be they in the Milky Way or in distant galaxies. (For the record, the star that is closest to the Sun is Proxima Centauri, in the Alpha Centauri system, 4.2 light years away.) In theory, on any one night it is possible to see up to 2,500 stars twinkling away when looking up at the dark sky from the Earth. And if you gaze at them though a pair of binoculars or a telescope, it should become apparent that not all are the bright, white colour that they typically appear when viewed with the naked eye. Their different colours are caused by their different temperatures, indicating that they are at different stages of their evolution. But before following the stellar life cycle from birth to death, it may be helpful to establish what is meant by a star's magnitude (brightness) and spectral type (temperature and colour), and how astronomers represent these values on a graph, thereby instantly enabling them to tell the star's age.

There are an estimated 200 billion stars within the Milky Way.

MAGNITUDES, SPECTRAL TYPES AND THE HERTZSPRUNG–RUSSELL DIAGRAM

The first person to devise a scale with which to classify stars by *apparent magnitude*, or their brightness when viewed from the Earth, was the Greek astronomer Hipparchus, over a hundred years before the birth of Christ. His scale ranged from one to six, but the scale that 21st-century astronomers use is far more extensive, and, unlike his, includes negative (minus) values. Indeed, the brightest stars are today assigned negative numbers. Sirius, or Alpha [α] Canis Majoris, for example, is the most brilliant star in the sky as seen from the Earth and has an apparent magnitude of –1.45. Far fainter stars are given positive numbers. The dimmest stars visible with the naked eye from the Earth, for example, have magnitudes of around +6. Each value signifies a two-and-a-half-times increase or decrease in brightness.

A star's *absolute magnitude* – that is, its real brightness, or *visual luminosity* – is its brightness in comparison to the Sun, or how the star would appear if it were 10 parsecs (32.6 light years) away from Earth. A light year is the equivalent of 9,500 trillion km (5,903 trillion miles). A star's mass determines its visual luminosity, and the greater the mass, the brighter the luminosity. (Note that a star's mass is measured in relation to that of the Sun: a high-mass star may have a mass as high as 50 solar masses, while a low-mass star may have a mass that is one-tenth of a solar mass. Another point to be aware of is that the brightness of non-stellar celestial objects, such as Venus and

The Hertzsprung-Russell diagram is a tool with which astronomers can plot a star's age and evolutionary status.

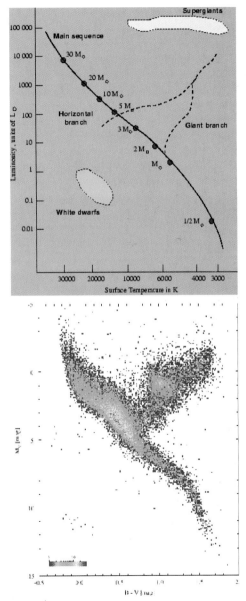

the Moon, which generate no light of their own, but instead reflect sunlight, can also be measured by magnitude.)

A star's colour is influenced by its temperature. The coolest stars are orangey-red in colour, and the hottest appear blue-white, for red and blue lie at opposite ends of the spectrum. Astronomers consequently divide stars into seven spectral-type categories according to their temperature, as follows:

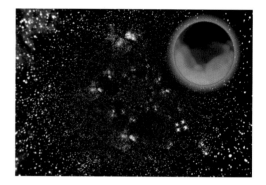

A computer-generated image of a star at the blue end of the spectrum.

- type O (the bluest stars): 40,000–29,000°C (72,032–52,232°F);
- type B: 28,000–9,700°C (50,432–17,492°F);
- type A: 9,600–7,200°C (17,312–12,992°F);
- type F: 7,100–5,800°C (12,812–10,472°F);
- type G: 5,700–4,700°C (10,292–8,492°F);
- type K: 4,600–3,300°C (8,312–5,972°F);
- type M (the reddest stars): 3,200–2,100°C (5,792–3,812°F).

Each spectral type is in turn divided into ten subsections numbered 0, 1, 2, 3, 4, 5, 6, 7, 8 and 9, with 0 representing the hottest value and 9, the coolest. Thus it is that the Sun is classified as having a G2 spectral type, while Sirius' spectral type is A0.

How do astronomers measure a star's brightness and temperature in order to determine its absolute magnitude and spectral type? The answer is that they use spectrometers, spectrographs, spectroscopes and other instruments of spectral analysis to capture the light that the star is generating, translate this starlight into wavelengths on an electromagnetic spectrum and then note where on the spectrum spectral lines appear. This tells them whether radiation is being absorbed or emitted by the star, and consequently more about its make-up and age, as well as how luminous and hot it is.

Certain stars are sometimes described as being 'main-sequence stars'. This description is derived from the Hertzsprung–Russell (H–R) diagram on which astronomers plot stars. In an H–R diagram, the x-axis, or horizontal axis, represents a star's spectral type (or surface temperature, or, alternatively, colour), while the y-axis, or vertical axis, signifies its visual luminosity (its absolute magnitude, or else apparent magnitude). Once a star's spectral type and visual luminosity, for instance, are known, its position is plotted on the graph. Astronomers can then tell at a glance what point it has reached in its life. Most stars fall into a band called the main sequence, which runs diagonally from the top left to the bottom right of the graph, the stars at the top left being hot, blue, large and high mass, and those at the bottom right being cool, red, small and low mass. These main-sequence stars are generating energy by means of nuclear reaction – that is, they are creating helium by burning hydrogen at their cores. Eventually, however, this activity will cease, and as they become increasingly cool and faint, they will merit being moved off the main sequence and into the region that indicates giant stars; there are also areas for protostars, supergiants and white dwarfs.

Read on to learn more about a star's life cycle.

The Sun is a star whose spectral type is G2.

A STAR'S LIFE CYCLE

A star can live for billions of years, a length of time that is almost impossible for the human brain to compute. Yet in some ways, what happens to the star during the course of its life mirrors the human experience, for all stars are born, mature, decline and eventually die.

New images from NASA's Spitzer Space Telescope allow us to peek behind the cosmic veil and pinpoint one of the most massive natal stars yet seen in our Milky Way galaxy. The never-before-seen star is 100,000 times as bright as the Sun. Also revealed for the first time is a powerful outflow of hot gas emanating from this star and bursting through a giant molecular cloud.

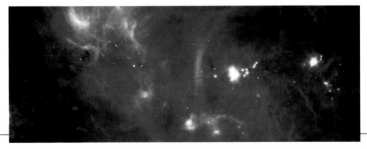

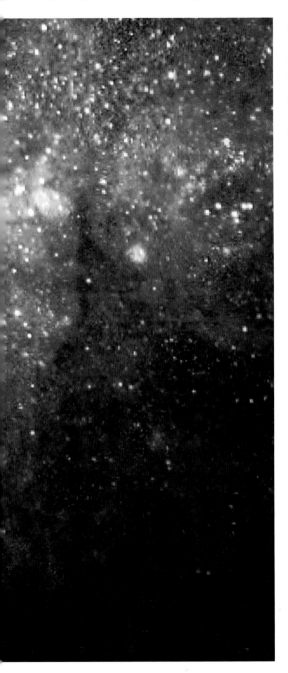

A star is born: protostars

Stars are born within giant molecular clouds (GMCs), cold, dark areas of the interstellar medium (see page 191) that are chock-full of dust and gas particles, and especially of the hydrogen molecules that are the prerequisite for stellar life. Stellar conception is triggered when a GMC's own gravity causes it increasingly to contract and heat up until it finally breaks up into clumps, or protostars. Cocooned within a cloud of gas and dust – a nebula that gradually assumes a disc shape as the protostar spins within it – each protostar continues to shrink, condense and become hotter at its core until it eventually begins to glow and emits gas from the edge of the surrounding disc. And when the temperature at its core hits the 10 million°C (18,000,032°F) mark, nuclear fusion, or the fusion of hydrogen nuclei to form helium, begins (see page 57 for a description of this process). This is the flashpoint that signals the birth of a star, entitling it to take its place on the main sequence (as a protostar, it would have appeared at the top right of an H–R diagram).

The nearby dwarf galaxy NGC 1569 is a hotbed of vigorous star-birth activity, which blows huge bubbles that riddle the galaxy's main body. This image was taken by a camera on the Hubble Space Telescope.

Main-sequence stars

Once nuclear fusion begins, the young star stops shrinking and enters a period of relative stability as it continues to convert hydrogen into helium. The higher its mass, the faster a star burns hydrogen, and consequently the greater its heat and luminosity. Its life will be considerably shorter than a low-mass star's, however, for no star enjoys an infinite supply of

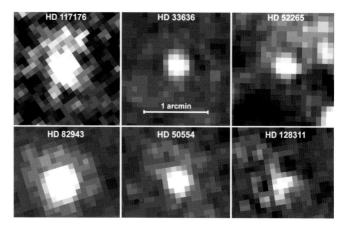

These stars are similar in age and temperature to our Sun. In astronomical terms, they are stellar main-sequence stars, with spectral types of F, G or K. These planet-bearing stars have a median age of 4 billion years. For reference, our Sun is classified as a G star, with an age of approximately 5 billion years.

hydrogen. Indeed, it is estimated that the lifespan of a star whose mass is the equivalent of ten solar masses is around 10 million years, while that of a star whose mass is the same as the Sun's is at least 10 billion years.

Dwarfs

Most stars are dwarfs, and of these, most are red dwarfs. A red dwarf is a main-sequence star whose mass is less than the Sun's, and that is consequently smaller, cooler and redder. Gliese 229 (whose alternative name is GL 229), in the constellation of Lupus, is an example of a red dwarf. Other dwarf stars may be divided into different colour categories, such as yellow dwarfs (including the Sun) and white dwarfs (see page 148). Brown dwarfs are very cool, faint, low-mass protostars that never began to burn hydrogen and can thus be considered failed stars. Neither brown nor white dwarfs are main-sequence stars.

Astronomers using NASA's Hubble Space Telescope have stumbled upon a mysterious object that is grudgingly yielding clues to its identity. A glance at the Hubble picture at top shows that this celestial body, called He2-90, looks like a young, dust-enshrouded star with narrow jets of material streaming from each side. In fact, the object is classified as a planetary nebula, the glowing remains of a dying, lightweight star. But the Hubble observations suggest that it may not fit that classification either. The Hubble astronomers now suspect that this enigmatic object may actually be a pair of ageing stars masquerading as a single youngster. One member of the duo is a bloated, red-giant star shedding matter from its outer layers. This matter is then gravitationally captured in a rotating, pancake-shaped accretion disc around a compact partner, which is most likely a young white dwarf (the collapsed remnant of a Sun-like star). The stars cannot be seen in the Hubble images because a lane of dust obscures them.

Variable stars

As their name suggests, variable stars are stars whose apparent magnitude does not remain constant – in other words, they appear to twinkle when observed from Earth, sometimes regularly, and sometimes irregularly. There are various categories of variable stars: eruptive, rotating, eclipsing, pulsating and cataclysmic, as well as Cepheids, or Cepheid variables.

An *eruptive variable* is a star whose apparent magnitude changes erratically due to events occurring around it. It may be, for example, that a stellar wind is blowing away excess

Variable stars appear to twinkle when viewed from Earth.

dust and gas from the surface of a young star, thereby temporarily shrouding its light from view. Eta [η] Carinae, whose magnitude varies from between –0.8 and +7.9, is another type of eruptive variable: believed to be a supergiant (see page 145), its variation in brightness is caused by the Homunculus Nebula that surrounds it.

A *rotating variable* is a star whose surface is unevenly covered with spots similar to sunspots (see page 59), so that as the star rotates, a constantly changing pattern of spots presents itself to Earthly viewers, causing the star's apparent magnitude to vary. Cor Caroli A, whose magnitude regularly varies from between +2.84 and +2.96 over the course of five-and-a-half days, is a rotating variable.

An *eclipsing variable* (or eclipsing binary, see page 153) is actually two stars which are so close to one another that not only do they appear as one, but when their orbits are edge-on to the Earth and one passes in front of the other from our Earthly viewpoint, the star that is furthest from us is eclipsed. If the star in question is bright, the dimming of its light will be striking, but if it is faint, the difference in its apparent magnitude will be less dramatic. Algol, or Beta [β] Persei, in the constellation of Perseus, is an example of a regularly eclipsing variable whose magnitude varies from between +2.1 and +3.4 every three days or so.

A *pulsating variable* is a star that has

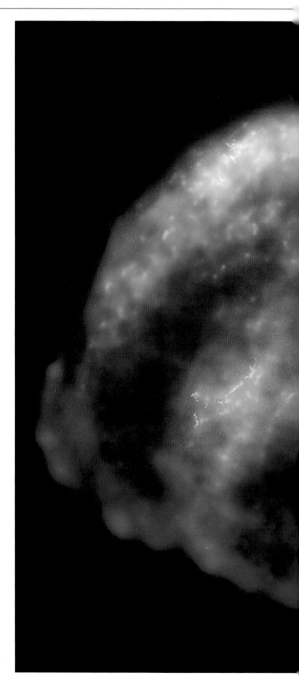

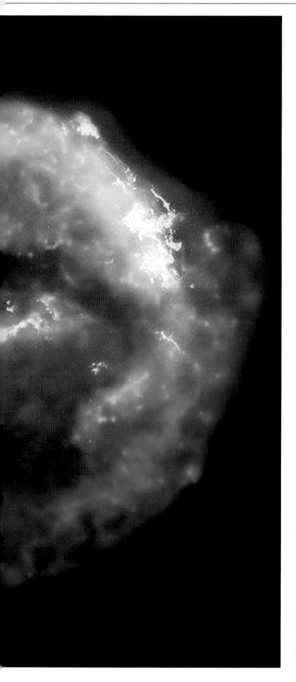

evolved into its red-giant or giant-star phase (see page 145), when its changing size and temperature cause its apparent magnitude to fluctuate. Mira (Omicron [o] Ceti) is a regularly pulsating variable whose magnitude ranges from between +2.0 and +10.1 over a period of 332 days, making it a long-period variable (LPV).

Cataclysmic variables include supernovae, which are supergiant stars that suddenly brighten spectacularly when they explode as a result of violent structural changes (type II supernovae). Another type of cataclysmic variable is called a cataclysmic binary, or a nova, i.e., the nuclear explosion caused when the build-up of hydrogen gas sucked by a white dwarf in a binary-star system (see page 153) from its companion star, a red dwarf, suddenly explodes; and a type Ia supernova occurs when the white dwarf itself collapses and explodes. (See page 149 for more on supernovae.) The first-recorded nova is thought be one that occurred near the constellation of Antares (Alpha [α] Scorpii) that was noted by Chinese star-gazers in 1300 BC. In 1975, astronomers recorded a nova in the constellation of Cygnus that increased the binary star's

NASA's three 'great observatories' – the Hubble Space Telescope, the Spitzer Space Telescope and the Chandra X-ray Observatory – joined forces to probe the expanding remains of a supernova, called Kepler's supernova remnant, first seen 400 years ago by sky-watchers, including astronomer Johannes Kepler.

apparent magnitude by 40 million times.

Cepheid variables have been of particular significance to astronomers since 1912, when the US astronomer Henrietta Leavitt noticed that their luminosity corresponds to the length of their cycles of variation (that is, the time that it takes for them to go from their brightest, to their dimmest, to their brightest again): the brighter a Cepheid variable, the longer its cycle of variation. She then realised that it was possible to work out their absolute magnitude and consequently to calculate their distance from the Earth. This was a significant discovery because until then, astronomers had been limited to the parallax (angular) method of measuring stellar distances. This was useless for really distant stars, which barely appear to move over a period of six months when viewed from an Earthly vantage point. Yellow supergiants whose spectral types are F and G, Cepheid variables now act as standard candles, enabling astronomers to calculate the distance from the Earth of the globular clusters and galaxies that contain them. (According to the inverse square law, if you know that two stars' absolute magnitudes are the same and one appears brighter, it must be closer.) Their namesake star is Delta [δ] Cephei, whose magnitude varies from between +3.5 and +4.4 over a regular period of nearly five-and-a-half days. It was discovered by the English astronomer John Goodricke (1764–86) in 1784.

A NOTE ON VARIABLE-STAR NAMES

Variable stars may be known by their traditional names, such as Antares, or their Bayer-system name, which, in the case of Antares, is Alpha Scorpii or α Scorpii. Alternatively, they may be designated with the letter 'V' and a number greater than 335, or else with a capital letter and a constellation name, for example, R Coronae Borealis (an eruptive variable) or T Coronae Borealis (a cataclysmic variable).

A red–giant star.

Giant stars

The store of hydrogen that fuels a low-mass, main-sequence star's nuclear reaction and the generation of helium at its core eventually starts to run low, and ultimately runs out. There is still some hydrogen in an outer shell, however, which the star starts to burn, causing it to expand – its diameter may grow from 10 to 100 times that of the Sun's – and its surface to cool, sometimes to as low as 3,600°C (6,512°F). Giant stars with a spectral type of K, like Aldebaran (Alpha [α] Tauri), in the constellation of Taurus, are called red giants. (Giants can be classified as being any colour from blue to red on the spectrum, depending on their surface temperature. Blue is the hottest, and red is the coolest.) Now burning helium at its core, and producing heavier elements like carbon and oxygen as by-products, a giant star is no longer a main-sequence star. Initially, it is moved to the right on an H–R diagram to reflect its falling temperature, and it is then shifted upwards to indicate its growing brightness.

Supergiant stars

Whereas low-mass, main-sequence stars evolve into giant stars, high-mass stars become supergiants, which, with diameters up to 1,000 times that of the Sun, are 10 times larger. Like giants, supergiants burn helium, but the carbon and oxygen that this process generates combine further to create even heavier elements, including iron, which builds up in the ever-hotter core. Also in common with giants, supergiants are classified according to their surface temperature and consequently colour. Rigel (Beta [β] Ori) is an example of a blue supergiant, while the enormous red supergiant Mu [μ] Cephei is known as the 'Garnet Star' on account of its hue.

NASA's Spitzer Space Telescope has captured in stunning detail the spidery filaments and newborn stars of the Tarantula Nebula, a dense, star-forming region also known as 30 Doradus. This cloud of glowing dust and gas is located in the Large Magellanic Cloud, the nearest galaxy to our own Milky Way, and is visible primarily from the southern hemisphere. This image of an interstellar cauldron provides a snapshot of the complex physical processes and chemistry that govern the birth – and death – of stars. At the heart of the nebula is a compact cluster of stars known as R136, which contains very massive and young stars. The brightest of these blue supergiant stars are up to 100 times more massive than the Sun, and are at least 100,000 times more luminous. These stars will live fast and die young, at least by astronomical standards, exhausting their nuclear fuel as they will in a few million years.

A DYING STAR IN THE ANT NEBULA

This false-colour image from NASA's Spitzer Space Telescope shows a dying star (centre) surrounded by a cloud of glowing gas and dust. Thanks to Spitzer's dust-piercing, infrared eyes, the new image also highlights a never-before-seen feature – a giant ring of material (red) slightly offset from the cloud's core. This clumpy ring consists of material that was expelled from the ageing star. The star and its cloud halo constitute a 'planetary nebula' called NGC 246. When a star like our own Sun begins to run out of fuel, its core shrinks and heats up, boiling off the star's outer layers. Leftover material shoots outwards, expanding in shells around the star. This ejected material is then bombarded with ultraviolet light from the central star's fiery surface, producing huge, glowing clouds – planetary nebulas – that look like giant jellyfish in space.

In this image, the expelled gases appear green, and the ring of expelled material appears red. Astronomers believe that the ring is probably made of hydrogen molecules that were ejected from the star in the form of atoms, then cooled to make hydrogen pairs. The new data will help explain how planetary nebulae take shape, and how they nourish future generations of stars.

This nebula was imaged on 20 July 1997 and 30 June 1998, by Hubble's Wide Field and Planetary Camera 2. The Ant Nebula, whose technical name is Mz3, resembles the head and thorax of an ant when observed with ground-based telescopes. The new Hubble image, with 10 times the resolution revealing 100 times more detail, shows the 'ant's' body as a pair of fiery lobes protruding from a dying, Sun-like star. The Ant Nebula is located between 3,000 and 6,000 light years from Earth in the southern constellation Norma.

The image challenges old ideas about what happens to dying stars. This observation, along with other pictures of various remnants of dying stars called planetary nebulae, shows that our Sun's fate will probably be much more interesting, complex and dramatic than astronomers previously believed.

Although the ejection of gas from the dying star in the Ant Nebula is violent, it does not show the chaos that one might expect from an ordinary explosion, but instead shows symmetrical patterns. One possibility is that the central star has a closely orbiting companion whose gravitational tidal forces shape the outflowing gas. A second possibility is that as the dying star spins, its strong magnetic fields are wound up into complex shapes like cooked spaghetti in an eggbeater. Electrically charged winds, much like those in our Sun's solar wind, but millions of times denser and moving at speeds up to 1,000 km per second (more than 600 miles per second) from the star, follow the twisted field lines on their way out into space.

Planetary nebulae and white dwarfs

Just as it ran out of hydrogen when it was a main-sequence star, so a red giant will one day exhaust its supply of helium. And when that happens, its core will shrink, but the rest of the giant star will expand, to the extent that its outer layers are suddenly blown away. Now detached from the star of which it once formed a part, the newly created planetary nebula, as this discarded shell of stardust and gas is somewhat misleadingly termed (planetary nebulae once being thought to be planetary discs), gradually disperses into space. Then all that is left of the red giant is a white dwarf, the star's white-hot, dense, Earth-sized core, whose mass is no greater than 1.4 times that of the Sun. It may take billions of years to die, but this is the star's last incarnation: having nothing left to burn, the white dwarf is fated to dwindle away into nothingness. (That said, if it is in a binary system, the white dwarf may eventually explode as a type I supernova, see page 150.) In its parallel life on an H–R diagram, the white dwarf exists at the bottom left, until it becomes too cold and faint to be registered anymore. But that is not quite the end. It may be almost undetectable, but a ghostly presence will remain imprinted on the fabric of time and space, namely a black dwarf, a sort of 'memory' of the blazing star that it once was.

During the millennia or so that they remain hanging around the white dwarfs at their centres, planetary nebulae are beautiful to behold. Illuminated by the white dwarf's light, a planetary nebula resembles an ethereal cloud when viewed from the Earth through a telescope. The powerful gravitational

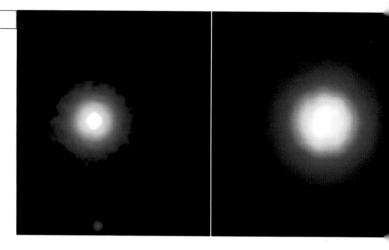

NASA's Spitzer Space Telescope recently captured these images of the star Vega, located 25 light years away in the constellation Lyra. Spitzer was able to detect the heat radiation from the cloud of dust around the star and found that the debris disc is much larger than was previously thought. This side-by-side comparison, taken by Spitzer's multi-band-imaging photometer, shows the warm, infrared glows from dust particles orbiting the star at wavelengths of 24 microns (left, in blue) and 70 microns (right, in red). Scientists compared the surface brightness of the disc in the infrared wavelengths to determine the temperature distribution of the disc and then infer the corresponding particle size in the disc. Most of the particles in the disc are only a few microns in size, or 100 times smaller than a grain of Earth sand.

field surrounding the incredibly dense white dwarf may also affect the planetary nebula's appearance. For example, some may form a simple shape, such as the Ring Nebula that lies south-east of Vega (Alpha [α] Lyrae), in the constellation of Lyra. Others' more complex shapes may seem oddly reminiscent of more mundane objects. Train a telescope on the constellation Vulpecula, for instance, and you may see the fuzzy, dumbbell-like outline of the Dumbbell Nebula; or focus on Ophiuchus, and you may be rewarded with the sight of the delicate lobes of Minkowski 2–9, a butterfly planetary nebula. If it is a white dwarf that you are interested in observing, look for Sirius B, which is currently orbiting Sirius (Alpha [α] Canis Majoris).

Supernovae and nebulae

A supergiant whose mass as a main-sequence star was more than eight solar masses will not go out with a whimper, but with a bang! When the iron that is steadily accumulating within its core reaches 1.4 solar masses – called the Chandrasekhar limit after Subrahmanyan Chandrasekhar (1910–95), the Indian astrophysicist who calculated it – the core implodes, triggering an almost incomprehensively enormous explosion callked a supernova. It may last only a few seconds, but the explosive flash is extraordinarily bright – indeed, with an estimated luminosity a billion times that of the Sun, it is brighter than all of the stars within a galaxy put together.

The Crab Nebula.

The collapsed core has now become either a tiny neutron star (see below) or a black hole (see page 193).

Further out in space, the explosive shockwave generated when the gaseous material that once enclosed the core is blasted away – at speeds of up to 20,000 km (12,428 miles) per second – heats the gas within the interstellar medium (see pages 191-3) to such high temperatures that it glows, illuminating the supernova remnant – now a nebula – as it continues to expand and disperse over the course of thousands of years. The Crab Nebula, for example, is the remnant of a supernova recorded by Chinese astronomers as occurring within the constellation of Taurus in 1054. The supernova itself was so bright that it was visible for three weeks during the daytime and for two years at night.

Supernovae are relatively rare occurrences (only five have been observed in the Milky Way over the past thousand years, the last by Johannes Kepler in 1604), which is why astronomers were excited to witness the blue supergiant Sanduleak –69°202 exploding in a nearby galaxy, the Large Magellanic Cloud (LMC), in 1987. Supernova (SN) 1987A, as it was officially dubbed, was a type II supernova caused by an exploding supergiant, as described above. There is another type of supernova, however, namely type I, of which there are three subtypes, Ia, Ib and Ic. Type Ia, for example, occurs when a white dwarf in a binary system sucks so much gas from its companion star that it cannot support its increased mass and explodes (see also cataclysmic variables, page 143). Because the magnitude generated by type Ia supernovae is always the same, astronomers use them as standard candles for measuring the distance to other galaxies from the Earth.

Neutron stars, pulsars and magnetars
Unless it becomes a black hole (see page 193), following a supernova, the collapsed core of a supergiant will survive as a neutron star, an incredibly dense body comprising a solid

iron crust containing the liquid neutrons formed by the fusion of protons and electrons. Although it measures only about 30 km (19 miles) in diameter, a neutron star's mass is roughly equivalent to that of the Sun, and its gravitational and magnetic fields are consequently incredibly powerful.

Some, like the neutron star at the centre of the Crab Nebula, in the constellation of Taurus, spin exceedingly fast. In the process, they throw out regular pulses of radiation – particularly radio waves – from their magnetic poles, which is why they are called pulsating stars, or pulsars. The first neutron star to be designated a pulsar was PSR 1919+21, which came to the attention of British astronomers Antony Hewish and Jocelyn Bell Burnell when they picked up its pulsing radio signal in 1967. PSR 1919+21 is known as a radio pulsar, but there are also optical pulsars, which visibly flash (such as the pulsating neutron star in the Crab Nebula), X-ray pulsars and gamma-ray pulsars (which emit X-rays and gamma rays respectively, although some, such as Geminga, give out both). A pulsar may also be termed a binary pulsar if it is orbiting another star. Magnetars, which have an even stronger magnetic field, are thought to be forms of neutron star, too.

Over millions of years, a spinning neutron star will gradually slow down until it is rotating too slowly to send out waves of electromagnetic radiation from its poles.

Resembling sparks from a fireworks display, this image taken by a Jet Propulsion Laboratory (JPL) camera onboard NASA's Hubble Space Telescope shows delicate filaments that are sheets of debris from a stellar explosion in the nearby Large Magellanic Cloud galaxy.

STAR CLUSTERS

Although the life cycle of a single star was outlined above, it is important to understand that stars are not born, and do not always live and die, in solitary splendour. If you consider that a giant molecular cloud disintegrates into clumps, and that each of these clumps develops into a protostar, and then a star, it makes sense that some stars appear to form a pair, a group of three or more, or even a cluster of hundreds, thousands or millions of stars, all bound together by the force of gravity.

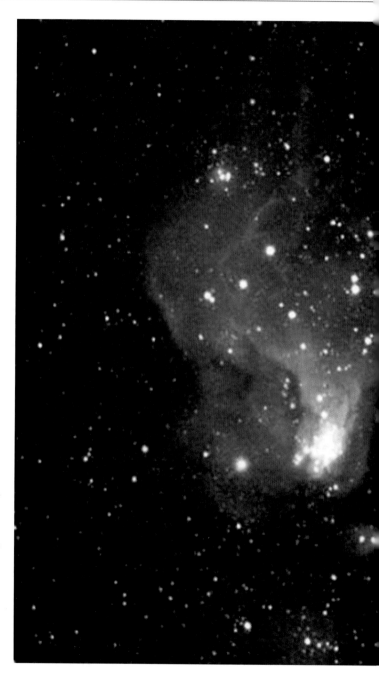

This glowing gas cloud, called Hubble-V, has a diameter of about 200 light years. A faint tail of gas and dust trailing off the top of the image sits opposite a dense cluster of bright stars at the bottom of the irregularly shaped nebula. The HST's resolution and ultraviolet sensitivity reveal a dense knot of dozens of ultra-hot stars nestled in the nebula. Each star glows 100,000 times brighter than our Sun. These 4-million-year-old stars, considered youthful in the cosmic time scale, are too distant and crowded together to be resolved from ground-based telescopes. The small, irregular host galaxy, called NGC 6822, is one of the Milky Way's closest neighbours. It lies 1.6 million light years away in the direction of the constellation Sagittarius.

The binary, or double-star, system

Two stars that have been captured within one another's gravitational field are called a binary or double star, or are together said to form a binary or double-star system. The two stars are orbiting around the same centre of balance, which, if they are of equal mass, lies at an equidistant point between them. If one star has a greater mass than the other, however, the centre of balance will be located nearer to that star. And the greater the distance between the two companion stars, the longer their orbital period.

There are various types of binary star, as follows.

- An *astrometric binary* is a double-star system in which only one star is visible, the other making itself apparent by affecting the visible star's proper motion (see page 160).

- A *visual binary* is a double-star system that comprises two distinctly separate stars.

- A *spectroscopic binary* consists of two stars that are so close that they appear to be one star, but are individually identifiable by their spectra. A spectroscopic binary may also be an *eclipsing binary*, or an eclipsing variable, see page 142.

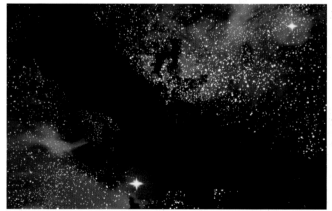

Double stars are two stars that gravity has forced to pair up. .

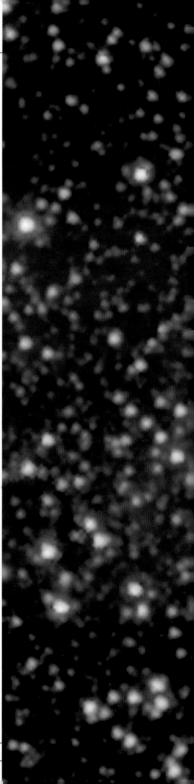

- An *interacting binary* is a double-star system in which the two stars are so close that they are exchanging gas.
- A *semi-detached* binary comprises a small, hot star and a larger, cooler star that is losing gas to its companion.
- In a *contact binary*, the two stars are so close that they are touching one another.

Note, however, that if two stars appear to be aligned when viewed from the Earth, they may be an *optical pair*, and not a true binary star, or physical double star.

Multiple stars

A multiple star, or multiple-star system, consists of three or more stars. In the case of three, two companion stars usually form a binary star, while the third orbits around them. In a double binary system, two pairs of binary stars orbit a shared centre of balance. Alpha [α] Centauri, in the constellation of Centaurus, is a multiple star comprising three stars; Epsilon [ε] Lyrae, in the constellation of Lyra, is an example of a double binary system; while Sigma [σ] Orionis, in the constellation of Orion, is made up of five stars.

Open and globular clusters

Star clusters can consist of tens, hundreds, thousands, and even hundreds of thousands, or millions, of stars. Open and globular clusters are of interest to astronomers partly because the stars within them came into being at around the same time, and partly because, in the case of globular clusters, it is likely that they were formed at the same time as their parent galaxy (and they are not confined to the Milky Way alone).

This false-colour image taken by NASA's Spitzer Space Telescope shows a globular cluster previously hidden in the dusty plane of our Milky Way galaxy. Globular clusters are compact bundles of old stars that date back to the birth of our galaxy, 13 or so billion years ago. Astronomers use these galactic 'fossils' as tools for studying the age and formation of the Milky Way.

A cluster of newborn stars herald their birth in this interstellar Valentine's Day commemorative picture obtained in ???? with NASA's Spitzer Space Telescope. These bright, young stars are found in a rosebud-shaped (and rose-coloured) nebulosity known as NGC 7129. The star cluster and its associated nebula are located at a distance of 3,300 light years in the constellation Cepheus. A recent census of the cluster reveals the presence of 130 young stars. The stars formed from a massive cloud of gas and dust that contains enough raw materials to create 1,000 Sun-like stars. In a process that astronomers still poorly understand, fragments of this molecular cloud became so cold and dense that they collapsed into stars. Most stars in our Milky Way galaxy are thought to form in such clusters.

Open clusters (which are also called galactic clusters) are made up of hundreds or thousands of loosely associated young stars, often surrounded by nebulae. (Astronomers know that the stars are young because the largest appear blue, which means that they are hot, and therefore young.) This type of irregularly shaped cluster is found in the Milky Way's spiral arms. An example of an open cluster is the Pleiades, in the constellation of Taurus, a group of around 100 stars, of which seven are visible with the naked eye; the Pleiades is thought to be 78 million years old and is 375 light years away. The stars within an open cluster will eventually disperse.

Globular clusters are massive conglomerations of hundreds of thousands, and sometimes millions, of old stars that, in our galaxy, are found around the galactic bulge and as far away as the galactic stellar halo, i.e., in the outer regions of the Milky Way. Because they are so densely packed (due to the gravitational attraction of the stars within them), they appear larger, brighter and more spherical than open clusters. Their largest stars, typically giants, being older and cooler than those in open clusters, appear yellow in colour. Three globular clusters are visible from the Earth with the naked eye, which perceives them as being a single, fuzzy star: 47 Tucanae, in the constellation of Tucana, which is 15,000 light years away; Omega [ω] Centauri, in the constellation of Centaurus, which is 17,000 light years away; and, at 23,000 light years away, M13 (or the Great Cluster), in the constellation of Hercules.

A NOTE ON CLUSTER NAMES

Star clusters can be called by a descriptive name, such as the evocative Beehive Cluster, an open cluster comprising around 50 stars in the constellation of Cancer that is also known as Praesepe. Star clusters may have a less poetic catalogue number, too. According to the Messier catalogue, for instance, the Pleiades is M45. A star cluster may also have a New General Catalogue (NGC) number and an Index Catalogues (IC) number.

THE CONSTELLATIONS

When our ancestors gazed up at the night sky, they discerned patterns of stars that they associated with mundane objects and creatures, which is how many of the constellations got their names. In some cases, the connection is clear: look at the constellation of Scorpio – in the sky or on a star map (see pages 224-8) – and the curl of a scorpion's tail is instantly apparent. Sometimes a myth is associated with the constellation. For example, a tale from ancient Greece describes how the giant hunter Orion boasted that he could slaughter any creature on Earth, whereupon the goddess Gaea, annoyed by his presumption, commanded a scorpion to sting him on his heel. The scorpion accordingly unleashed the sting in its tail and killed Orion, who was later restored to life by

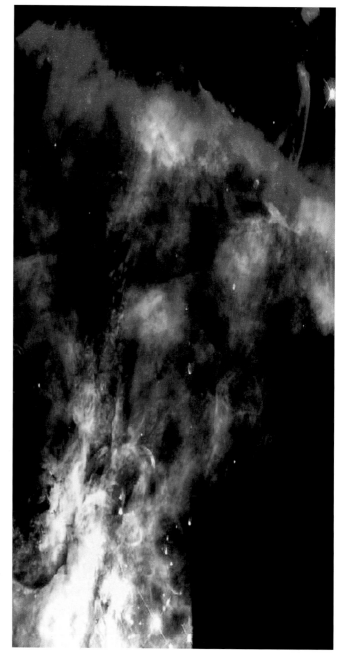

The constellation Orion.

How Orion became a constellation

Orion was the son of Neptune. He was a handsome giant and a mighty hunter. His father gave him the power of wading through the depths of the sea, or, as others say, of walking on its surface.

[Orion] dwelt as a hunter with Diana, with whom he was a favourite, and it is even said she was about to marry him. Her brother was highly displeased and often chid her, but to no purpose. One day, observing Orion wading through the sea with his head just above the water, Apollo pointed it out to his sister and maintained that she could not hit that black thing on the sea. The archer-goddess discharged a shaft with fatal aim. The waves rolled the dead body of Orion to the land, and bewailing her fatal error with many tears, Diana placed him among the stars, where he appears as a giant, with a girdle, sword, lion's skin, and club. Sirius, his dog, follows him, and the Pleiads [sic] fly before him.

Thomas Bulfinch,
The Age of Fable, or Stories of Gods and Heroes
(1855)

Asclepius, who stamped on the scorpion and killed it. You can read a description of how Orion himself is said to have become a constellation on page 159. If you browse through a book on classical mythology, you will find many such links, for the ancient Greeks and Romans believed that the gods conferred immortality on their favourites by literally 'placing them among the stars'.

In around AD 150, the Greek astronomer Ptolemy drew up a list of 48 constellations, which was added to over the hundreds of years that followed. Not all of their names are drawn from mythology, and some 'new' southern-hemisphere constellations, such as Triangulum Australe, or the Southern Triangle, are both admirably descriptive and succinct. Today, there are officially 88 constellations. Since 1930, the word 'constellation' technically no longer applies to a star group, but to an area of sky that contains a constellation. See pages 224–32 for further information on the star maps that chart constellations, as well as the constellations that you should be able to observe, depending on whether you are in the northern or southern hemisphere.

The visual connection between some constellations and their namesakes is not always clear, for which there is sometimes a simple explanation. One of the definitions of the word 'constellation' is that it signifies one of 88 groups of bright stars as seen from the Earth and the solar system. It may not be apparent to us, but stars move under the influence of the gravitational force exerted by other celestial objects in their galaxy. And because stars move (and astronomers can measure their 'proper motion', as it is called, by recording the changes in their spectra caused by the Doppler effect, see page 47), the constellations have changed in shape since they were first named and recorded.

The zodiac

If you are aware of your sign of the zodiac, you may be interested to know that astronomy was once synonymous with astrology. The 12 signs of the zodiac are Aries, Taurus, Gemini, Cancer, Leo, Virgo, Libra, Scorpio, Sagittarius, Capricorn, Aquarius and Pisces. Also termed zodiacal constellations, they lie along the ecliptic, or the ecliptic plane. This is an imaginary path, inclined at an angle of 23.45° to the celestial equator, along which the Sun, Moon and planets travel during the course of a year, so that they appear to move through each zodiacal constellation in turn. Astronomically speaking, there are also further zodiacal constellations, but these are disregarded in astrology.

In astrology, Aries is the first sign of the zodiac and kicks off the zodiacal year when the Sun is said to enter it on, or around, 21 March. This is also the approximate date of the vernal equinox – the start of spring in the northern hemisphere and of autumn in the southern hemisphere – when the Sun did once indeed enter the constellation of Aries, which is why the vernal equinox is also called the first point of Aries. In fact, due to the precession of the equinoxes, the Sun now crosses the celestial equator in Pisces. Similarly, although the autumnal equinox, which occurs on, or around 23 September, when the Sun crosses the celestial equator again six months later, is called the first point of Libra, the Sun is actually still in the constellation of Virgo on this date.

THE ZODIACAL CONSTELLATIONS

Aries, the Ram.

Taurus, the Bull.

Gemini, the Twins.

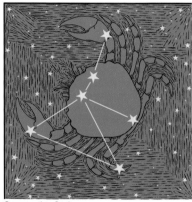

Cancer, the Crab.

Leo, the Lion.

Virgo, the Virgin.

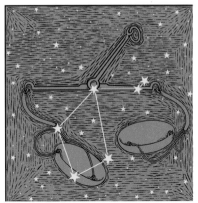

Libra, the Scales.

Scorpio, the Scorpion.

Sagittarius, the Archer.

Capricorn, the Sea Goat.

Aquarius, the Water Carrier.

Pisces, the Fishes.

A NOTE ON CONSTELLATION AND STAR NAMES

Different cultures have called the constellations different names. Ursa Major, for example, is known as the Great Bear or the Plough in Britain, where it was once also dubbed Charles' Wain, while it is called the Big Dipper in the USA, the Großer Wagen (Great Wagon) in Germany and the Rishis, or Sages, in India, to name but a few of its traditional names. To avoid confusion, the International Astronomical Union has assigned each of the constellations an official, Latin name and a three-letter abbreviation, as detailed in the table on pages 166-9. Informally, as well as in an astrological context, the constellations (and particularly the 12 for which the signs of the zodiac are named) may be called by their common names, or the English translation of their formal, Latin names, so that Sagittarius may be known as 'the Archer', for example.

The individual stars within the constellations can similarly have many names. Take Sirius (a Latin name derived from the Greek word seirios, or 'scorching'), or the Dog Star, in the constellation of Canis Major, for example, which the ancient Egyptians knew as the goddess Sopdet, or Sothis, and the Chinese, as Thien Lang, or the Celestial Wolf. Sirius has an apparent magnitude of −1.45, an actual magnitude of +1.5, and its spectral type is A0 (a white star). As the brightest star in the sky above the Earth, it has been included in most star catalogues, and consequently has many numerical designations, including 48915 in the Henry Draper (HD) Catalogue and 151881 in the Smithsonian Astrophysical Observatory Star Catalogue. However, it is generally known either by its traditional name (Sirius) or by its Bayer-letter-system name. In the Bayer system, a star's apparent magnitude is indicated by a Greek letter (see page 169 for the Greek alphabet), starting with alpha for the brightest, followed by the genitive, or possessive, form of its constellation name. When the Greek letter is spelled out, both it and the 'of [whatever constellation]' part of its name are capitalised, but the symbol for the letter may be used instead, and the genitive of the constellation name abbreviated. In this way, Sirius becomes Alpha Canis Majoris, α Canis Majoris or α CMa. Another popular naming system is the Flamsteed number system, in which a number precedes the genitive of the constellation name according to the star's position in the constellation in order

Canis Major, the Great Dog.

of right ascension from west to east, so that Sirius becomes 9 Canis Majoris. And just to complicate matters further, because Sirius is actually a binary star, the star that we can see shining brightly is now designated Sirius A, while its companion star, a fainter white dwarf, is identified as Sirius B.

THE CONSTELLATIONS
(listed in alphabetical order)

Latin name	Genitive	Abbreviation	Common name
Andromeda	Andromedae	And	Andromeda
Antlia	Antliae	Ant	The Air Pump
Apus	Apodis	Aps	The Bird of Paradise
Aquarius	Aquarii	Aqr	The Water Carrier
Aquila	Aquilae	Aql	The Eagle
Ara	Arae	Ara	The Altar
Aries	Arietis	Ari	The Ram
Auriga	Aurigae	Aur	The Charioteer
Boötes	Boötis	Boo	The Herdsman
Caelum	Caeli	Cae	The Chisel
Camelopardalis	Camelopardalis	Cam	The Giraffe
Cancer	Cancri	Cnc	The Crab
Canes Venatici	Canum Venaticorum	CVn	The Hunting Dogs
Canis Major	Canis Majoris	CMa	The Great Dog
Canis Minor	Canis Minoris	CMi	The Little Dog
Capricornus	Capricorni	Cap	The Sea Goat (or Goat Fish)
Carina	Carinae	Car	The Keel
Cassiopeia	Cassiopeiae	Cas	Cassiopeia
Centaurus	Centauri	Cen	The Centaur
Cepheus	Cephei	Cep	Cepheus
Cetus	Ceti	Cet	The Whale
Chamaeleon	Chamaeleontis	Cha	The Chameleon
Circinus	Circini	Cir	The Compass
Columba	Columbae	Col	The Dove
Coma Berenices	Comae Berenicis	Com	Berenice's Hair
Corona Australis	Coronae Australis	CrA	The Southern Crown
Corona Borealis	Coronae Borealis	CrB	The Northern Crown
Corvus	Corvi	Crv	The Crow

Latin name	Genitive	Abbreviation	Common name
Crater	Crateris	Crt	The Cup
Crux	Crucis	Cru	The Southern Cross
Cygnus	Cygni	Cyg	The Swan
Delphinus	Delphini	Del	The Dolphin
Dorado	Doradus	Dor	The Dorado (or Swordfish)
Draco	Draconis	Dra	The Dragon
Equuleus	Equulei	Equ	The Foal
Eridanus	Eridani	Eri	The River Eridanus
Fornax	Fornacis	For	The Furnace
Gemini	Geminorum	Gem	The Twins
Grus	Gruis	Gru	The Crane
Hercules	Herculis	Her	Hercules
Horologium	Horologii	Hor	The Clock (or Pendulum Clock)
Hydra	Hydrae	Hya	The Water Snake
Hydrus	Hydri	Hyi	The Little (or Lesser) Water Snake
Indus	Indi	Ind	The Indian
Lacerta	Lacertae	Lac	The Lizard
Leo	Leonis	Leo	The Lion
Leo Minor	Leonis Minoris	LMi	The Little Lion
Lepus	Leporis	Lep	The Hare
Libra	Librae	Lib	The Scales
Lupus	Lupi	Lup	The Wolf
Lynx	Lyncis	Lyn	The Lynx
Lyra	Lyrae	Lyr	The Lyre
Mensa	Mensae	Men	The Table (or Table Mountain)
Microscopium	Microscopii	Mic	The Microscope

Latin name	Genitive	Abbreviation	Common name
Monoceros	Monocerotis	Mon	The Unicorn
Musca	Muscae	Mus	The Fly
Norma	Normae	Nor	The Level (or Square)
Octans	Octantis	Oct	The Octant
Ophiuchus	Ophiuchi	Oph	The Serpent Bearer
Orion	Orionis	Ori	Orion (or the Hunter)
Pavo	Pavonis	Pav	The Peacock
Pegasus	Pegasi	Peg	Pegasus (or the Winged Horse)
Perseus	Persei	Per	Perseus
Phoenix	Phoenicis	Phe	The Phoenix
Pictor	Pictoris	Pic	The Easel (or Painter's Easel)
Pisces	Piscium	Psc	The Fishes
Piscis Austrinus	Piscis Austrini	PsA	The Southern Fish
Puppis	Puppis	Pup	The Stern
Pyxis	Pyxidis	Pyx	The Compass (or Mariner's Compass)
Reticulum	Reticuli	Ret	The Net
Sagitta	Sagittae	Sge	The Arrow
Sagittarius	Sagittarii	Sgr	The Archer
Scorpius	Scorpii	Sco	The Scorpion
Sculptor	Sculptoris	Scl	The Sculptor
Scutum	Scuti	Sct	The Shield
Serpens	Serpentis	Ser	The Serpent
Sextans	Sextantis	Sex	The Sextant
Taurus	Tauri	Tau	The Bull

Latin name	Genitive	Abbreviation	Common name
Telescopium	Telescopii	Tel	The Telescope
Triangulum	Trianguli	Tri	The Triangle
Triangulum Australe	Trianguli Australis	TrA	The Southern Triangle
Tucana	Tucanae	Tuc	The Toucan
Ursa Major	Ursae Majoris	UMa	The Great Bear (or Big Dipper)
Ursa Minor	Ursae Minoris	UMi	The Little Bear (or Little Dipper)
Vela	Velorum	Vel	The Sails
Virgo	Virginis	Vir	The Virgin
Volans	Volantis	Vol	The Flying Fish
Vulpecula	Vulpeculae	Vul	The Fox (or Little Fox)

THE LETTERS OF THE GREEK ALPHABET

Name	Symbol	Name	Symbol
Alpha	α	Nu	ν
Beta	β	Xi	ξ
Gamma	γ	Omicron	o
Delta	δ	Pi	π
Epsilon	ε	Rho	ρ
Zeta	ζ	Sigma	σ
Eta	η	Tau	τ
Theta	θ	Upsilon	υ
Iota	ι	Phi	φ
Kappa	κ	Chi	χ
Lambda	λ	Psi	ψ
Mu	μ	Omega	ω

The Milky Way
and Other Galaxies

We thought differently not even a century ago, but we now know that the Earth is just one of the planets orbiting the Sun, and that the Sun is only one of around 200 billion stars within the Milky Way, which in turn is merely one of the estimated 100 billion galaxies (or 'island universes', as they were once called) that the universe contains.

It was Edwin Hubble who opened up a whole new area of cosmological research when, in 1924, as a result of photographing Cepheid variables in the constellation of Andromeda, he concluded that the fuzzy-looking object labelled M31 in the Messier Catalogue was, in fact, another galaxy, and not part of the Milky Way, as had previously been thought.

This image is a Galaxy Evolution Explorer observation of the large galaxy in Andromeda, Messier 31, or M31. The Andromeda Galaxy is the most massive in the Local Group of galaxies that includes our Milky Way. Andromeda is the nearest large galaxy to our own.

A spiral galaxy.

When calculating the distances of certain galaxies from our own, astronomers continue to use Cepheid variables, as well as type Ia supernovae, as standard candles (see pages 144 and 150). And while astronomers can use parallax to measure the distances of stars, and employ radar to calculate the distances of planets, neither method is effective beyond the bounds of the Milky Way, when they have to resort to alternative means. One way of measuring a spiral galaxy's distance, for instance, is by employing the galaxy-rotation

Scientists are seeing unprecedented detail in the spiral arms and dust clouds of the nearby Whirlpool Galaxy, thanks to a new Hubble Space Telescope image. The image uses data collected on 15 and 24 January 1995 and 21 July 1999 by Hubble's Wide Field and Planetary Camera 2. Using the image, a research group led by Dr Nick Scoville, of the California Institute of Technology, Pasadena, clearly defined the structure of the galaxy's cold dust clouds and hot hydrogen, and they linked star clusters within the galaxy to their parent dust clouds.

method, in which astronomers use spectrometers to measure the red- and blueshifts (see page 47) on each side of a spiral galaxy in order to ascertain how fast it is spinning. If the galaxy is rotating at the same speed as one that contains Cepheid variables, whose distance can be worked out, their distances from the Earth are likely to be similar.

At around 2.5 million light years away, M31, or the Andromeda Galaxy, as it is now commonly called, is the most distant celestial object visible from the Earth with the naked eye. As a result, it marks the limit of our 'observable universe'. Other galaxies are so far away that even when the most powerful optical instruments are brought into play, they remain invisible to us. Our knowledge of them is, in many cases, consequently almost completely derived from spectral analysis. That said, the Hubble Space Telescope (HST) delivered some real eye-openers when it sent images back to the Earth in 1995 of galaxies thought to be 10 billion light years away. Indeed, the technological sophistication inherent in 21st-century observatories is rapidly enabling us to learn more about galaxies: about their types, about their contents, and about the clusters and superclusters that they form, as well as about what lies between them. But let us start our survey of the galactic universe with the galaxy that we know best: our own.

A NOTE ON GALAXY NAMES

When astronomers refer to a particular galaxy, they may use its common name, such as the Andromeda Galaxy; its Messier number, in this instance, M31; its New General Catalogue (NGC) number, Andromeda's being NGC 224; or its Index Catalogues (IC) number. Note that when distinguished with an upper-case 'G', 'Galaxy' alone refers to the Milky Way.

Jupiter . . . summoned the gods to council. They obeyed the call, and took the road to the palace of heaven. The road, which anyone may see in a clear night, stretches across the face of the sky, and is called the Milky Way. Along the road stand the palaces of the illustrious gods; the common people of the skies live apart, on either side.

Thomas Bulfinch,
The Age of Fable, or Stories of Gods and Heroes (1855)

Astronomers using NASA's Hubble Space Telescope have found a spiral galaxy that may rotate in the opposite direction to what was expected.

THE MILKY WAY

The ancient Greeks and Romans were not the only people who believed that the Milky Way was a road that traversed the heavens; other cultural traditions – those of Egypt, India and China, for instance – tell of it being a celestial river. And if you are lucky enough to be looking up at the night sky from a spot that is unclouded by light pollution, and can therefore admire the milky-looking band of starlight arcing from horizon to horizon that is an Earthling's view of the Milky Way, you'll understand how these beliefs arose. Today, however, we generally accept that the Milky Way and its fellow galaxies came into being following the big bang, when the gravitational pull of the dark matter within the universe (see page 194) tugged thousands of gaseous clouds together, ultimately causing stars to form and their own, increasing force of gravity to become powerful enough to hold them together.

The Milky Way is a barred-spiral (SBbc) galaxy (see page 183) that rather resembles an aircraft, or the 'flying saucer' type of UFO, in profile, and a spinning Catherine wheel when viewed head-on. At its centre is a flattened sphere, or spheroid, called the galactic bulge, which is about 6,000 light years thick. When seen in profile, a flatter 'wing', or arm, about 2,000 light years thick, tapers out from either side of the galactic bulge. The whole of this galactic structure measures about 100,000 light years across. Alternatively, if you imagine that you were looking down on the Milky Way from above, it would resemble a circle (the

The Milky Way.

hub-like galactic bulge), with a number of spiral arms coiling around it. This galactic disc is surrounded by a star-studded, galactic stellar halo that is, in turn, enclosed by a dark galactic halo, a sort of ghostly remnant of the Milky Way's circumference when it began life as a ball of gas. The Milky Way's brightest, or major, arms are the Sagittarius Arm (the inner arm), the Perseus Arm (the outer arm) and the Orion Arm, or the Local Arm (which lies between the Sagittarius and Perseus arms). From measuring the wavelengths emitted by neutral hydrogen atoms in the galactic disc with radio telescopes, astronomers have calculated that the galactic disc is rotating at a speed of

220 km (137 miles) per second, and that the stellar halo is moving more slowly, at an estimated 50 km (31 miles) per second.

The Sun, which the Earth is, of course, orbiting, is a relatively insignificant star that is positioned about 25,000 light years (two-thirds of the way) from the galactic centre, on the Orion Arm, so that if you imagine the Milky Way in profile again, we are looking at the galactic disc edge-on. This is why the stars that are crowded into this area appear to us as though they are arranged in a belt, with those in the galactic stellar halo being scattered less densely on either side. Dust obscures the stars in the galactic bulge from our view, however, apart from those that can be glimpsed through 'Baade's Window', a clear spot in the constellation of Sagittarius, where the Milky Way – which looks its brightest between June and September – consequently appears at its widest and most densely packed.

In summary, the Milky Way's make-up is as follows.

● At the heart of our rotating, barred-spiral galaxy is a radio source named Sagittarius A*, which is believed to be a supermassive black hole (see page 193).

● The galactic bulge surrounding this black hole comprises about 10 billion yellow and red, cool, old stars and very little gas and dust.

● The spiral arms contain dark, molecular clouds, glowing nebulae (see pages 147 and 149) and billions more blue-white, hot, young stars, often grouped into open clusters.

● The galactic stellar halo holds an estimated 200 globular clusters, as well as some individual stars, all separated, and enclosed, by dark matter.

GALACTIC CLASSIFICATIONS

The Milky Way is a spiral galaxy, which is a common galactic classification, as are elliptical galaxies. From the HST's focus on an area of space called the Hubble Deep Field, we know that most galaxies were created by the collision of gas clouds, and that spiral galaxies resulted if these clouds were revolving around one another before fusing, while elliptical galaxies came into being if the clouds were not whirling around when they merged. Not only has the HST revealed this process of galaxy formation, but it has provided evidence that large galaxies can engulf smaller ones in this way, too.

Elliptical galaxies include both the largest and smallest galaxies (the latter being termed 'dwarf ellipticals', or 'dwarf spheroidals') in the universe, but spiral galaxies are usually comparable in size. Galaxies may furthermore be classified as being lenticular, barred spiral or irregular.

HUBBLE'S 'TUNING-FORK' DIAGRAM

In 1925, Edwin Hubble devised a diagram in the shape of a tuning fork to help astronomers to classify galaxies into categories. In the diagram, each category and subcategory is assigned a label, as follows:

- elliptical (E) galaxies: E0 to E7, ranging in shape from the roundest to the flattest oval; spheroidal (Sph) and dwarf spheroidal (dSph) galaxies were discovered later, and are considered to be related to, if not subcategories of, the elliptical group;
- lenticular galaxies: S0;
- spiral (S) galaxies: Sa, Sb and Sc (to which has since been added another category, Sd), with Sa having a large galactic bulge and tightly coiled, spiral arms, this structure 'relaxing' through Sb and Sc, with Sd displaying the smallest galactic bulge and loosest spiral arms;
- barred-spiral (SB) galaxies: SBa, SBb and SBc, with the 'a', 'b' and 'c' designations having identical meanings to those associated with spiral (S) galaxies; and
- irregular (Irr) galaxies: Irr-I galaxies contain gas, nebulae and young stars, while Irr-II galaxies do not, or at least do not appear to; there are also dwarf irregulars (dIrr).

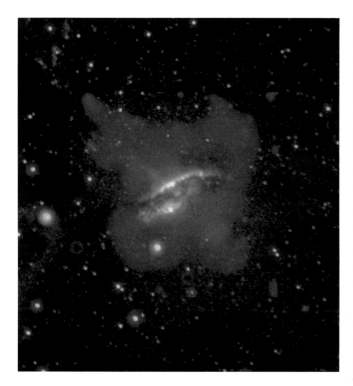

This image of the active, or radio, galaxy Centaurus A was taken by NASA's Galaxy Evolution Explorer on 7 June 2003. The galaxy is located 13 million light years from Earth and is seen edge-on, with a prominent dust lane across the major axis.

Elliptical galaxies

Because they no longer contain much dust or gas, ingredients that are vital for star birth, and because the stars within them are generally red – in other words, elderly, in stellar terms – elliptical (E) galaxies are thought to be among the oldest in the universe. They can be circular or oval in shape, do not have spiral arms and hardly rotate, if they rotate at all.

M105, in the constellation of Leo, is an E0-category galaxy that is thought to be situated 38 million light years from the Milky Way, while M59, in the constellation of Virgo, is a class-E5 galaxy that is calculated to lie 50 million light years away.

Lenticular galaxies

Lenticular (S0), or lens-shaped, galaxies combine elements that are characteristic of elliptical galaxies with others that are more typical of spiral galaxies, in that they comprise a galactic bulge containing relatively old stars, which is surrounded by a galactic disc that encompasses younger stars. Lenticular galaxies are probably closer to ellipticals than spirals, however, for they lack spiral arms and exhibit a large proportion of old stars, but little dust and gas.

NGC 5866, or the Spindle Galaxy, in the constellation of Draco, is a lenticular galaxy with an estimated distance of 35 million light years from the Milky Way.

Spiral galaxies

Spiral (S) galaxies comprise an elliptically shaped galactic bulge populated by older, reddish stars, cocooned by a flatter galactic disc made up of at least two spiral arms that are home to younger, blue-white stars, nebulae and significant quantities of gas and dust. The spirals that the younger, brighter stars describe are caused by the compacting action of the 'density wave' generated by the spiral galaxy's rate, and hence force, of rotation.

The Sombrero Galaxy, in the constellation of Virgo, is a category-Sa galaxy that is believed to lie 40 million light years away; the Andromeda Galaxy is an example of an Sb galaxy; and the Triangulum Galaxy (in the constellation of the same name) is an Sc galaxy around 2.5 million light years away.

The magnificent, spiral arms of the nearby galaxy Messier 81 (M81) are highlighted in this NASA Spitzer Space Telescope image. Located in the northern constellation of Ursa Major (which also includes the Big Dipper, or Great Bear), this galaxy is easily visible through binoculars or a small telescope. M81 is located at a distance of 12 million light years from Earth.

The Sombrero Galaxy is an Sa galaxy.

The double cluster NGC 1850 lies in a neighbouring satellite galaxy, the Large Magellanic Cloud, which is categorised as an irregular galaxy.

Barred-spiral galaxies

Barred-spiral (SB) galaxies differ from spiral galaxies only in that they have a 'bar' of stars running through the galactic bulge and extending to either side of this central hub, as described for the Milky Way above.

NGC 2859, in Leo Minor, is an SBa galaxy about 72 million light years from the Milky Way; NGC 5850, which is around 100 million light years away and can be found in the constellation of Virgo, is an SBb-category galaxy; while an example of an SBc galaxy is NGC 7479 in Pegasus, around 110 million light years away. Note that the Milky Way is classified as being an SBbc galaxy because the size of its galactic bulge and the looseness of its spiral arms fall somewhere between those respectively signified by the labels SBb and SBc.

Irregular galaxies

As the name that unites them suggests, irregular galaxies do not conform to a regular or uniform shape. They contain gas and dust, and consequently nebulae and young stars, but are generally relatively small and faint in comparison to other galaxy categories.

Irregular (Irr) galaxies include the Large Magellanic Cloud (LMC), in the constellation of Dorado, which is around 160,000 light years from the Milky Way.

Scientists using NASA's Hubble Space Telescope are studying the colours of star clusters to determine the age and history of starburst galaxies, a technique somewhat similar to the process of learning the age of a tree by counting its rings. This Hubble Heritage image showcases the galaxy NGC 3310. It is one of several starburst galaxies, which are hotbeds of star-formation, being studied by Dr Gerhardt Meurer and a team of scientists at Johns Hopkins University in Laurel, Maryland, USA.

SOME OTHER GALAXY TYPES

Although all galaxies are classified as being elliptical (or spheroidal), lenticular, spiral, barred spiral or irregular, they can also be said to belong to certain subtypes, notably starburst galaxies and the various types of active galaxies.

Starburst galaxies

When two galaxies that are moving at a rapid rate smash into one another, evidence of this galactic collision may be apparent in the form of a 'starburst', or an area within a

galaxy that appears crammed with hot, bright, young stars, indicating that they were born together out of a violent event. Such a pattern can be seen in the middle of the Cartwheel Galaxy, a once classic-looking spiral or lenticular galaxy about 500 million light years away in the constellation of Sculptor, whose starburst is thought to have been caused by a smaller galaxy pushing through its centre. The HST, as well as the Infra-red Astronomical Satellite (IRAS), has provided evidence that such collisions are occurring all of the time. Astronomers believe that a similar fate will befall the Milky Way and the Andromeda Galaxy in about 5 billion years' time. A galaxy may also be transformed into a starburst galaxy by a nearby galaxy's tidal, or gravitational, pull rather than a collision.

Active galaxies

Active galaxies include Seyfert galaxies, radio galaxies, quasars and blazars. They have earned their 'active' designation because they contain active galactic nuclei (AGNs) and emit enormous amounts of electromagnetic radiation, or energy, from a relatively tiny region at their centre. In all instances, it is thought that the ultimate source of this unusually concentrated energy emission is a supermassive black hole (see page 193), or an 'engine' that is topped by a 'distributor', namely an accretion disc that absorbs and emits the radiation in the form of polar or lateral jets. A torus (an enclosing, and sometimes obscuring, ring) of dust and gas may alter the observer's perception of the active galaxy, so that it may appear as a Seyfert or radio galaxy, or else as a quasar or blazar.

Seyfert galaxies

Seyfert galaxies are named after their discoverer, the US astronomer Carl Seyfert (1911–60), whose attention was first drawn to these spiral and barred-spiral galaxies with

This image, taken by NASA's Spitzer Space Telescope, shows in unprecedented detail the galaxy Centaurus A's last big meal: a spiral galaxy seemingly twisted into a parallelogram-shaped structure of dust. Spitzer's ability both to see dust and see through it allowed the telescope to peer into the centre of Centaurus A and capture this galactic remnant as never before. An elliptical galaxy located 13 million light years from Earth, Centaurus A is one of the brightest sources of radio waves in the sky. These radio waves indicate the presence of a supermassive black hole, which may be 'feeding' off the leftover galactic meal.

unusually bright centres in 1943. Astronomers now believe that all spiral galaxies will evolve into Seyfert galaxies, which currently include M77, in the constellation of Cetus, 45 million light years away, and NGC 4151, in the constellation of Canes Venatici, around 65 million light years away from us.

Radio galaxies

Radio galaxies are typically elliptical galaxies that emit radio waves up to 1 million times more powerful than those generated by ordinary galaxies. The first scientist to pick up evidence of a radio galaxy was the British physicist Stanley Hey (1909–2000), who detected radio waves emanating from the constellation of Cygnus in 1946. It was, however, the US

astronomers Walter Baade (1893–1960) and Rudolph
Minkowski (1895–1976) who, in 1954, tracked the source of
these radio waves to a galaxy now known as the Cygnus A
radio galaxy, 740 million light years from Earth.

Other examples of radio galaxies include Centaurus A, in
the constellation of Centarus, 13 million light years from the
Earth, and M87, which lies 50 million light years away, in
the constellation of Virgo.

Quasars

In 1963, when focusing on the radio source 3C 273, which is
positioned 2.1 billion light years away in the constellation of
Virgo, the Dutch astronomer Maarten Schmidt (b. 1929)
noted its dramatic redshift and concluded that it was the
furthest body from the Earth ever detected up till then. In
quasars, the torus is tilted, enabling us to see the light
emitted by the accretion disc, and their consequent
resemblance to faint stars prompted their appellation 'quasi-
stellar objects', QSOs, or 'quasars' for short.

Other quasars include 3C 48, which is situated 4.5 billion
light years away in the constellation of Triangulum, and PKS
2349-01, in the constellation of Pisces, 1.5 billion million
light years away.

Blazars

The radio waves picked up on Earth from BL Lacertae, 900
million light years away, in the constellation of Lacertra, in
1968, led to it becoming the first active galaxy to be
classified as a blazar. Blazars resemble quasars, except that
their luminosity varies dramatically, which is why they were
once believed to be variable stars (see page 141).

OJ 287, which is calculated to lie 3.8 billion light years
away in the constellation of Cancer, and 3C 279, 5.8 billion
light years away in the constellation of Virgo, have both
been identified as blazars.

WHAT DO GALAXIES CONTAIN?

If you've browsed through the preceding pages, you'll
already know that galaxies contain planets, moons,
meteoroids, asteroids and comets (see pages 54–131).
Astronomers have long wondered whether other solar
systems exist, and were consequently thrilled to discover a
planet orbiting the star 51 Pegasi in 1995. More of such
extrasolar planets, or exoplanets, have since been pinpointed,
and the large size – comparable to Jupiter's – and proximity

This artist's concept depicts a distant, hypothetical solar system, similar in age to our own. Looking inwards from the system's outer fringes, a ring of dusty debris can be seen, and within it, planets circling a star the size of our Sun. This debris is all that remains of the planet-forming disc from which the planets evolved. Planets are formed when dusty material in a large disc surrounding a young star clumps together. Leftover material is eventually blown out by solar wind or pushed out by gravitational nteractions with planets. Billions of years later, only an outer disc of debris remains.

to their parent stars exhibited by some have surprised scientists, as have the extremely elliptical orbits of others.

Galaxies also contain stars, of course, along with brown dwarfs, or 'failed stars' (for more detailed information on stars, see pages 134–69). Note that in the galactic context, stars are often said to fall into one of two categories: Population I stars and Population II stars. Population I stars range in age from very young to quite old, but all are packed with elements that are heavier than hydrogen and helium;

this population is generally confined to the spiral arms of spiral and barred-spiral galaxies. By contrast, Population II stars are both extremely old and contain very few heavy elements; this population of stars is typically found in the galactic bulge and galactic stellar halo.

Stars are separated from one another by the interstellar medium, which contains their birthplaces, giant molecular clouds (GMCs), and which also morphs in places into protoplanetary discs and nebulae; it carries intergalactic messages in the form of cosmic rays, gravitational waves and neutrinos. Black holes and dark matter are further fundamental, if mysterious, galactic components about which astronomers are striving to learn more.

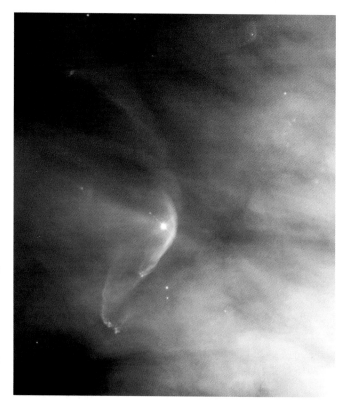

Astronomers using NASA's Hubble Space Telescope have found a bow shock around a very young star in the nearby Orion Nebula, an intense star-forming region of gas and dust. Named for the crescent-shaped wave that a ship makes as it moves through water, a bow shock can form in space when two gas streams collide. In this case, the young star, LL Ori, emits a vigorous wind, a stream of charged particles moving rapidly outwards from the star. Our own Sun has a less energetic version of this wind that is responsible for auroral displays on the Earth.

The interstellar medium

The matter that lies between the stars is called the interstellar medium (ISM). This consists primarily of molecular, neutral or ionised hydrogen and tiny amounts of cosmic-dust particles – icy, sooty traces of graphite and silicate that find their way into the ISM from the surfaces of old, cool stars and that make it appear cloudy in parts. It is estimated that 90 per cent of the Milky Way's mass is made up of stars and other celestial objects, the ISM contributing 5 per cent and molecular clouds, the remaining 5 per cent.

The ISM is neither static nor cool, but is constantly shifting and changing as stars are born and die, altering its temperature and make-up in the process. The warm intercloud medium, which is made up of areas of hydrogen gas, may glow pinkly at temperatures of around 8,000°C (14,432°F), for example. However, gas bubbles within it – including those that form the Cygnus Loop, which is a supernova remnant (SNR), or the visible after-effect of a supernova – may be dramatically hotter, while dense, star-creating and light-obscuring molecular and giant molecular clouds may be far cooler, at around –254°C (or –425°F).

Look towards the constellation of Cygnus, and you may see the Cygnus Rift, a dark band in the centre of the Milky Way that is actually a GMC, or a dark nebula. While protostars and stars are thought to be forming within GMCs, it is planets that are suspected of taking shape within protoplanetary discs, or proplyds, murky-looking spheres comprising 99 per cent gas and 1 per cent dust that are visible in the Orion Nebula, for instance. A protoplanetary disc cocoons a newborn, central star and it is the matter within it that slowly clumps together to create planetesimals, which later develop into protoplanets and eventually become planets (see page 54).

Other types of clouds that are detectable within the ISM are cool H I regions, which comprise neutral hydrogen atoms and emit radio waves, and warmer H II regions: areas of ionised hydrogen that are visible as the nebulae enveloping young stars. Indeed, a nebula's composition may be similar to that of a molecular cloud. The difference is that rather than being dark, the nebula is lit up by reflected starlight (when it is called a reflection nebula) or by the ultraviolet light emitted by the nebula from energy absorbed from newborn stars (when it is termed an emission nebula). The Orion Nebula, for instance, which lies 1,500 light years away in the constellation of Orion, is an emission nebula that is energised by the ultraviolet

radiation emanating from a cluster of young stars collectively named the Trapezium for the shape that it describes below Orion's Belt. (See pages 147-50 for more information on planetary nebulae and nebulae.)

The ISM additionally contains weak magnetic fields, whose effect on the particles of cosmic dust within it can make the ISM appear striped. And that's not all that can be found within the ISM. Astronomers are now focusing on learning more about the cosmic rays within it. These are invisible protons, or the cosmic radiation generated, it is thought, by quasars, supernovae and other explosive events. Its gravitational waves, or ripples in the fabric of space, are

Although we know that they exist, we can currently only speculate about the exact nature of black holes.

similarly believed to have been generated by cataclysmic cosmic occurrences, while its neutrinos, or particles with hardly any mass and no charge, are said to have arisen from the hottest of cosmic events, including supernovae and even the big bang.

Black holes

Although their existence had long been predicted – by the US physicist J. Robert Oppenheimer (1904–67) in 1939, among others – it was not until 1971 that the first black hole was identified by NASA's *Uhuru* satellite, namely Cygnux X-1, which has 16 solar masses. And although, in 1997, astronomers finally found proof that there is a supermassive black hole at the centre of the Milky Way, far more work still needs to be done before we can claim to understand these galactic enigmas fully.

Black holes are so called because they are areas of utter blackness from which nothing, not even light, can escape. Unless it has become a neutron star (which typically occurs when its core is lighter than three solar masses), a stellar black hole is all that remains after a supergiant star's core has collapsed, generating a cataclysmic supernova. This stellar remnant is now so dense that nothing – and in the universe, nothing travels faster than light – can resist its gravitational pull, so that if anything comes close enough, it will be sucked into oblivion.

Astronomers' knowledge of black holes is still in its infancy, and these objects are, of course, invisible, but we are nevertheless able to study their warping effects on the fabric of space-time, along with their influence on any companion stars. As a result, the current thinking is that a singularity (a point at which matter is infinitely compressed to infinitesimal volume) lies at the heart of a black hole, encompassed by a boundary called an 'event horizon'. The event horizon marks the Schwarzschild radius, which was

named for its discoverer, the German physicist Karl Schwarzschild (1873–1916). Effectively the edge of the black hole's gravitational well, this is the point of no return, or an irresistible curvature in space-time that prevents the gas pulled off a companion star, for instance, by the black hole's gravitational attraction, from breaking free. Instead, as it is pulled inwards and downwards in an angular spiral, the gaseous matter builds up to form a spinning accretion disc that, as it becomes hotter and hotter through the action of gravitational friction, emits electromagnetic radiation in the form of X-rays and gamma rays.

Supermassive, or galactic, black holes are believed to be the 'engines' that power active galaxies (see page 185). They are too gigantic, it is thought, to have been created by the collapse of a supergiant star, which is why astronomers speculate that their origin may have been the cataclysmic collapse of a massive cloud of gas.

Dark matter
The dark matter that is believed to enclose the Milky Way and other galaxies, thereby separating them from one another, remains something of a mystery to astronomers. Called dark matter because it is invisible, we can nevertheless detect its presence from its gravitational pull on galaxies and light rays. The present belief is that dark matter is made up of weakly interacting massive particles (WIMPs); neutrinos (see page 193); and the massive compact halo objects (MACHOs) that are located within a galaxy's outer halo, and which are thought to include brown dwarfs and black holes.

Dark matter, it is thought, comprises
WIMPs, neutrinos and MACHOs.

GALAXY CLUSTERS

Just as stars may be bound together by the force of gravity so that they form binary and multiple stars, as well as open and globular clusters, so individual galaxies may similarly be clustered together into groups (of around fewer than 50) and clusters. Galaxy clusters are categorised as being irregular or regular, with the former comprising mainly spiral galaxies, and the later, primarily elliptical ones, that have merged with one another. Some galaxy clusters – known as giant elliptical galaxies – are dominated by a central, giant galaxy, which was probably created when one galaxy 'swallowed' a smaller one. Whatever their exact type, all galaxy clusters emit X-rays and radio waves from their hot, central regions.

The Milky Way is just one of around 30-plus galaxies that together make up the galaxy cluster called the Local Group. The Local Group is dominated by two

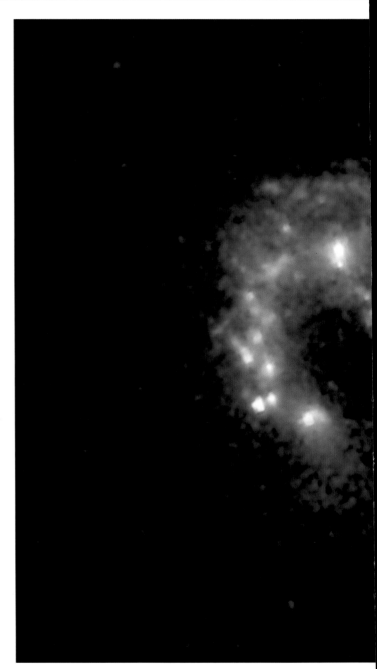

spiral galaxies: the Milky Way and the Andromeda Galaxy. These are surrounded by dwarf galaxies, including the Sagittarius Galaxy, which the Milky Way's powerful gravitational force is slowly, but inexorably, ripping apart. Two further notable members of the Local Group are the Large Magellanic Cloud (LMC) and the Small Magellanic Cloud (SMC), large, irregular galaxies that act as the Milky Way's satellite galaxies and that complete one full, elliptical orbit of it every 1.5 billion years. At approximately 160,000 light years away, the LMC is the closest large galaxy to ours (apart from the Sagittarius Galaxy). The SMC orbits the Milky Way at a distance of 190,000 light years. Eventually, it is speculated, both will end up being destroyed and consumed by the Milky Way.

This false-colour image from NASA's Spitzer Space Telescope reveals hidden populations of newborn stars at the heart of the colliding 'Antennae' galaxies. These two galaxies, known individually as NGC 4038 and 4039, are located around 68 million light years away, and have been merging for about the last 800 million years. The latest Spitzer observations provide a snapshot of the tremendous burst of star-formation triggered in the process of this collision, particularly at the site where the two galaxies overlap.

THE MAIN LOCAL GROUP GALAXIES
listed in order of distance from the Milky Way

Name	Type	Distance in light years	Diameter in light years
Milky Way	SBbc	0	100,000
Sagittarius	E/dSph	78,000	15,000
LMC	Irr	160,000	30,000
SMC	Irr	190,000	20,000
Ursa Minor	E/dSph	225,000	1,000
Draco	E/dSph	248,000	500
Sculptor	E/dSph	250,000	1,000
Carina	E/dSph	280,000	500
Sextans	E/dSph	290,000	1,000
Fornax	E/dSph	430,000	3,000
Leo II	E/dSph	750,000	500
Leo I	E/dSph	880,000	1,000
Phoenix	Irr/dSph	1,270,000	1,000
NGC 6822 (Barnard's)	Irr	1,750,000	8,000
And II	E/dSph	1,910,000	2,000
NGC 147	E/Sph	1,920,000	10,000
NGC 185	E/Sph	2,000,000	6,000
Andromeda	Sb	2,500,000	150,000
M32	E2	2,500,000	5,000
NGC 205	E/Sph	2,500,000	10,000
Triangulum	Sc	2,500,000	40,000
IC 1613	Irr	2,500,000	12,000
LGS 3 (Pisces Dwarf)	Irr/dSph	2,500,000	1,000
And I	E/dSph	2,570,000	2,000
And III	E/dSph	2,570,000	3,000
EGB0427+63	E/dSph	2,600,000	1,000
Tucana	E/dSph	2,900,000	500
WLM	Irr	3,000,000	7,000
SagDIG	Irr	3,700,000	5,000
IC 10	Irr	4,000,000	6,000
Pegasus	Irr/dSph	5,800,000	7,000

At approximately 50 million light years away, the galaxy cluster nearest to the Local Group is the irregular Virgo Cluster, which lies in the region of the constellation of Virgo and counts three giant elliptical galaxies and innumerable spiral galaxies among its estimated 2,000 members. Another notable neighbouring galaxy cluster is the regular Coma Cluster – which describes the form of a sphere and has a diameter of 20 million light years – that is thought to be positioned approximately 300 million light years away in the direction of the constellation Coma Berenice. The Coma Cluster is thought to comprise a minimum of 1,000 elliptical and lenticular galaxies, and maybe as many as 3,000.

Astronomers have ascertained that the distance between galaxy clusters is growing, a finding that both supports the big-bang theory and the premise that the universe is expanding (see page 47).

GALAXY SUPERCLUSTERS

When galaxy clusters merge, they become galaxy superclusters, or the largest-known structures in the universe, measuring as they do around 100 million light years across. These galaxy superclusters are arranged in the form of filaments, or strings, and are separated by unimaginably huge voids.

Our Local Group is a member of the Local Supercluster, a cluster consisting of 11 prominent 'clouds', at whose centre lies the Virgo Cluster, while the Coma Cluster is positioned at the heart of the Coma Supercluster.

Reach for the Stars!

A nebula in the constellation Cygnus.

If learning something about the history and theory of astronomy has whetted your appetite, how about taking the first practical steps towards becoming an astronomer and experiencing the reality for yourself? This chapter prepares you to embark on a fascinating, life-enhancing journey of discovery, which need not be expensive, requiring as it does only a little interest, enthusiasm and commitment.

DO YOUR HOMEWORK

Gazing up at the twinkling stars will always be a magical experience, but you'll find it far more exciting if you know a little about what you are looking at. Do even the tiniest bit of homework, and you'll reap immensely satisfying rewards.

If you have the opportunity to visit a planetarium, seize it! Reading the written word can be very instructive, but visual demonstrations are often far easier to grasp. The same goes for the relevant section of your local science museum. Don't forget to record any programmes on astronomy and space that are screened on television – you'll find building up a visual reference library in this way invaluable.

Do yourself a favour and start to familiarise yourself with the terms that astronomers use (you'll find plenty in the

A digitally generated image of a constellation.

glossary (see page 234), for starters. You may find them difficult to understand at first, but they'll soon become second nature, and it won't be long before you're fluent in the language of astronomy. Also learn about the 88 constellations (see pages 158–69): memorise their names, as well as those of the principal stars that they contain, and try to commit their appearance to memory, too.

As well as dipping into reference books and surfing the Internet to visit astronomy websites (see page 256 for a few interesting ones), buy an astronomy magazine once in a while. These contain the latest astronomical news, tips on what to look out for and when, as well as illuminating articles and inspiring photographs. Also consider joining an astronomy club.

Acquaint yourself with the language and symbols of astronomy.

DAYTIME OBSERVATIONS

If you only have limited opportunities to observe the sky at night, there are plenty of phenomena to look out for during the day. A sunset – and a sunrise, too – can be a truly spectacular sight, for example. And if you're wondering what causes the Sun, which usually appears yellow (although it is actually white) to give the impression of dissolving into such a stunning mixture of yellows, oranges and reds as it sinks below the horizon, the answer is simple. When the Sun is low in the sky, its light must pass through more of the Earth's atmosphere than when it is high, and thus through a greater number of atmospheric molecules. Colours at the blue end of the spectrum, which have short wavelengths, are scattered and absorbed by molecules in the Earth's atmosphere, so that they appear to disappear completely at sunset, while

The Sun sets over the drowned forests of Lake Kariba, Zimbabwe.

the yellow, orange and red colours at the opposite end of the spectrum pass straight through, and consequently display themselves to us in their full, fiery glory.

Observing solar phenomena

It is vital that you understand that you must never look directly at the Sun. If you want to observe it, you should instead project its image onto a piece of card, as described on page 208. Once you have mastered preparing this set-up, you could record changes in the Sun's sunspots by attaching a fresh piece of paper to your screen each day and then marking the position of any sunspots and faculae that appear on it with a pencil. (And don't forget to date each piece of paper to ensure that your daily records remain in the correct order.) Note that you will probably have to ask someone to hold the optic for you, or will otherwise somehow need to stabilise it so that it doesn't wobble, thereby creating a false image.

For more information on observing the Sun and solar eclipses, see pages 59–65.

Don't let the blotting-out of the Sun lull you into a false sense of security, and never look at the Sun directly.

SOLAR OBSERVATION: SAFEGUARD YOUR EYESIGHT!

Never, ever look directly at the Sun, not even while wearing a pair of sunglasses, and emphatically not while looking through a pair of binoculars or a telescope. If you were to do so – even for just a second or two – you would run the risk of blinding yourself. Permanently. Although Sun filters are sold for telescopes, be warned that these can overheat and then suddenly crack, which is why your best bet is to steer clear of them.

If you want to observe a partial, total or annular solar eclipse for yourself, the safest way to do so is to project the Sun's image onto a piece of card, as follows, and then to watch events unfolding on that.

YOU WILL NEED
● Either a pair of binoculars, a small piece of card, a pencil, scissors, and sticky tape;
● or a small refracting telescope with less than 30x magnification and an objective of under 100 mm (4 in) and, if necessary, a finder cap;
● a large piece of white card.

1. If you are using a pair of binoculars, place one objective (i.e., one of the lenses opposite an eyepiece) on the small piece of card and draw around it with a pencil. Now cut out the circle that you have traced and stick it over the lens with sticky tape.
2. If you are using a telescope, attach the finder cap.
3. Prop up the large piece of white card against a stable, vertical surface to act as a screen.
4. Hold the binoculars or telescope at a distance of about 50 cm (20 in) from the card, with the objective pointing towards the Sun. Move the binoculars or telescope around until a circular disc of light – an image of the solar disc – is projected onto the card. Focus the binoculars or telescope until you have as sharp an image as possible.

ENHANCING YOUR VISION

Unless it is near, eclipsing or transiting, the Sun, when you must take the precautions outlined on page 208, you can safely gaze at the Moon, as well as at any visible planet or star. There's a lot that you can see with the naked eye at night – the Moon, theoretically up to 2,500 stars and some of the planets, which will appear a little brighter than the stars – but far more will be revealed if you peer through a pair of binoculars or a telescope.

You can generally safely look at the Moon without taking any special precautions, but you will discern more detail using binoculars or a telescope.

This section briefly discusses the different types of optical instruments that are currently available to amateur astronomers. For more detailed advice, read an article or publication dedicated to the subject, or else consult a specialist vendor. But remember that it is not advisable to spend a lot of money on a pair of binoculars or a telescope until you are certain that you have been well and truly bitten by the astronomy bug, and your existing optical aids have proved themselves inadequate.

Binoculars

Effectively a pair of low-magnification telescopes that have been harnessed together, binoculars have the advantage of being more portable than telescopes, are often more comfortable to use and are usually cheaper.

The advantage of using binoculars over naked-eye observing is that they will enable you to see individual stars within the Milky Way and will give you a more detailed view of nebulae, star clusters and galaxies. If you are carrying out your observations in a place that is blighted by light pollution, a greater number of stars should become apparent when you raise your binoculars to your eyes.

You should find a set of numbers inscribed on a pair of binoculars: 7x50, 8x21, 8x30 or 10x50, for example. The first number represents the power of magnification (courtesy of the eyepiece lenses), and the second, the diameter of the objective lenses, or the lenses that collect light, so that 8x21 signifies a magnification of 8 and an objective-lens diameter of 21 mm (about ¾ in). Aim for a maximum magnification of only 10 – unless you're prepared to buy a tripod mounting – but the largest-diameter objective lens on offer, or that you can afford (the greater the diameter, the brighter the image). Don't be tempted by a pair of variable-power, or zoom, binoculars, however, because these may result in a distorted image. The accompanying literature should also specify the binoculars'

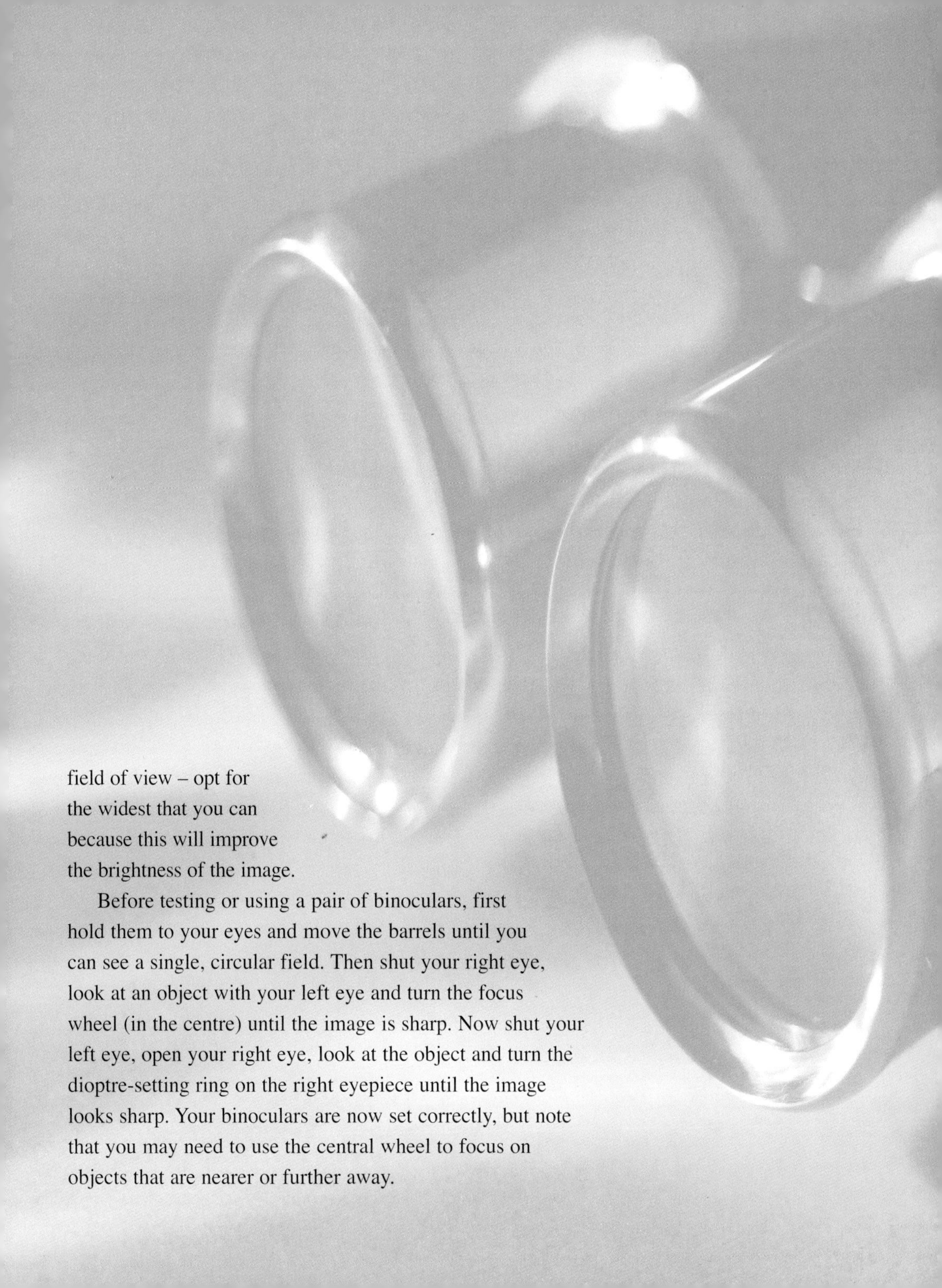

field of view – opt for
the widest that you can
because this will improve
the brightness of the image.

Before testing or using a pair of binoculars, first
hold them to your eyes and move the barrels until you
can see a single, circular field. Then shut your right eye,
look at an object with your left eye and turn the focus
wheel (in the centre) until the image is sharp. Now shut your
left eye, open your right eye, look at the object and turn the
dioptre-setting ring on the right eyepiece until the image
looks sharp. Your binoculars are now set correctly, but note
that you may need to use the central wheel to focus on
objects that are nearer or further away.

Telescopes

A telescope will usually give you a more detailed view of the stars than a pair of binoculars, but can be an expensive option, particularly if you succumb to temptation (or a salesperson's persuasive patter) and start splashing out on the many add-ons and accessories – filters, computerised object-locators and the like – that are available. Generally, the more expensive the telescope, the larger the aperture of the objective lens, the more light that it is able to gather as a result, and the better the image of the stars.

There are many subcategories, but you ultimately face a choice between two types of telescope: a refractor or a reflector. Their differences are explained in greater detail on pages 23–4, but the following information may be of help if you find yourself paralysed with indecision.

(And don't forget that unless you are really serious about astronomy, a pair of binoculars will often prove more than satisfactory for your purposes.)

Refracting telescopes
The basic components of a refracting telescope are a large objective lens that gathers and focuses light by refraction, the tube down which that light travels and the eyepiece lens that magnifies the light to give you an image. Unless it includes a corrective lens, be warned that the image produced will be upside down, the downside of the additional lens being a dimmer image, however. For the beginner, astronomers generally recommend that a refracting telescope's objective lens has a minimum diameter of around 76 mm (about 3 in) and is ideally one that is achromatic, or colour-corrected, to reduce the problem of false colour that is inherent in basic refracting telescopes. A finder, or finder scope, will often be attached to the tube; this is a low-magnification telescope whose sole purpose, as its name suggests, is to find an object on which you wish to focus.

Because it offers a clear contrast between light and dark areas, a refracting telescope may be a good choice if you are especially interested in observing the surface of the Moon or planets.

Reflecting telescopes
A reflecting telescope – or Newtonian reflector – differs from a refractor in that the light is gathered and reflected up the tube by a concave mirror at the bottom, rather than being refracted from the top by an objective lens; the image (which will appear upside down) is then reflected to a magnifying eyepiece in the side by a flat mirror. Dobsonian telescopes – forms of reflectors – are often suitable for beginners, but don't consider one with a mirror whose diameter is less than around 150 mm (approximately 6 in).

The advantage of a Newtonian reflector over a refractor is that the aperture is larger, so that fainter objects will be more visible, and there will be no false colour, but then a disadvantage is that the contrast will usually be less pronounced.

Hybrid telescopes

Some telescopes can be said to be refractor–reflector hybrids – catadioptric telescopes, for instance, which comprise both lenses and mirrors. The Maksutov or Schmidt-Cassegrain telescopes (SCT) are good, albeit expensive, options, largely because their tubes are shorter than that of a Newtonian reflector. They are also sold with a wide range of gizmos and accessories.

Mountings and cameras

Whatever type of telescope you opt for, you'll need a mounting to keep it steady, and consequently also the image that it produces. You again have two choices: an altazimuth mounting, which will enable you to move your telescope in any direction, or an equatorial mounting, which will allow you to move the telescope from side to side, or from west to east and vice versa.

A mounting, or tripod, is also essential if you intend to photograph celestial objects. A single-lens reflex (SLR) camera, a lens with a wide aperture and a film with a slow speed will usually give the best results (unless you are photographing meteors and comets, when a fast film is required).

You will need a tripod to keep your telescope steady.

SOME PRACTICAL TIPS

If it is misty, the Moon is full or you are in a town or city whose streets are illuminated by lights – in other words, an area that suffers from light pollution – it probably goes without saying that you will not be able to see as much as you would on a clear night, when the Moon is new or if you are in the middle of nowhere. There'll always be something to see, however, and the better prepared you are, the more you'll get out of the experience.

In the absence of the Sun's warmth, you can rapidly become very cold when standing outside in the cold and dark, so make sure that you are dressed warmly. Gloves are particularly important because frozen fingers are butter fingers that could damage a pair of binoculars or telescope by dropping them.

Most professional observatories are positioned in remote locations to avoid the problem of light pollution.

Here are some suggestions regarding other items that you may find useful to have to hand while star-gazing:

- some sheets of paper or, better still, a large notebook, and pencils;
- a watch;
- your binoculars or telescope;
- a stool to sit on, and maybe a folding table to put your equipment on;
- astronomy reference books, a planisphere or a star map (see pages 223–8);
- a torch whose light you've covered with red cellophane. It takes up to half an hour for your eyes to become accustomed to the dark, and covering your torch in this way will ensure that your slowly developed night vision isn't affected when you consult a star map, for instance.

PUTTING IT DOWN IN BLACK AND WHITE

There are lots of reasons why it's sensible to ensure that you have a good supply of paper and a pencil or two to hand:

- if you make a drawing of the Moon's appearance, you can compare it with a Moon map to see if you spotted any obvious lunar features; these become more visible as the lunar month progresses (for further details on viewing the Moon and lunar eclipses, see pages 76–84);
- if you witness a meteor shower and note down at what time it began, as well as the direction from which the meteors came, you could contribute your findings to one of the many organisations that study this phenomenon (see also pages 120–2).

Whatever it is that you are recording, don't forget to note down the time and date on which you made your sketch, along with your location and any optical instruments that you used.

AEROPLANE OR STAR?

Don't become too excited by a strange-looking light in the sky because it may be one of the following artificial objects and phenomena.

● A trail of light that is visible in the air around sunset is likely to be an aircraft's vapour trail (or condensation trail or contrail), which consists of water vapour from the engine that has frozen to form ice crystals, that has been illuminated by the fading sunlight.

● A bright flash in the sky while it is still light may be caused by a mobile-phone, or iridium, satellite's mirror catching the sunlight as it orbits Earth.

● A central light with a light on either side and one behind it, all moving quickly and smoothly in unison across the night sky, or else hovering in the air if you're near an airport, is likely to emanate from an aircraft.

● A light that appears larger than a star and is slowly, but visibly, moving across the sky is probably a large artificial satellite in orbit around the Earth.

Similarly, before calling a newspaper to report a UFO sighting, always ask yourself if the unusual object that you've spotted in the sky could have an equally logical, mundane explanation.

THE ASTRONOMICAL YEAR

The Earth orbits the Sun in around 365 days, which means that the night sky exhibits different sights, depending on where you are and the time of year.

The Sun's position in the sky appears to change during an Earthly year.

Because the Earth is tilted, during our planet's year-long

orbit, the height of the Sun in the sky gradually changes as it moves along the ecliptic (an imaginary path angled at 23.45° to the celestial equator close to which the Moon and planets also appear), resulting in the four seasons. These occur at different times in the two hemispheres into which the Earth is divided. The equator is the dividing line and the horizon is the line that marks the lower limit of what we can see. On the summer solstice, 21 June, the Sun is at its highest over the northern hemisphere, where it is summer; in the southern hemisphere, it is winter. On the winter solstice, which falls on 21 or 22 December, the Sun is at its highest over the southern hemisphere, where it is now summer; in the northern hemisphere, it is winter. And on the equinoxes – 21 March and 23 September – when the Sun crosses the Earth's equator, day and night are of equal length in both hemispheres. It is spring in the northern hemisphere and autumn in the southern hemisphere on the vernal equinox (21 March), while it is spring in the southern hemisphere and autumn in the northern hemisphere on the autumnal equinox (23 September). (See page 161 for more information on the Sun's passage along the ecliptic, the equinoxes and the zodiacal constellations.)

Our planet is also spinning on its axis at an angle of 23.5° from the vertical, and at a rate of once every 24 hours, which means that for 12 hours a day, each hemisphere in turn faces away from the Sun, when it is plunged into darkness, it becomes night and the stars become visible.

The Moon exhibits eight phases during the course of a lunar month.

It is the Earth that is rotating, while the stars' positions remain relatively fixed, but because we cannot register our planet's movement, the stars appear to be moving from east to west – the opposite direction to that in which the Earth is turning. The angle at which the Earth rotates barely alters, so that the North Pole is always pointing towards the Pole Star, or Polaris (in the constellation of Ursa Minor), while the South Pole remains in line with the Southern Cross, or the constellation of Crux. Such circumpolar constellations or stars (other stars are called equatorial constellations) never rise and set, although their position does change. This is also why there are certain constellations that are never visible in the northern or southern skies. If you are in the northern hemisphere, the Southern Cross will always be hidden from your sight, for instance, while if you are in the southern hemisphere, you will never see Polaris. But if you are at the equator, you will be able to see all of the constellations rising and setting throughout the course of a year.

In summary, if you don't live in a polar region, different stars will appear to rise and set throughout the year, and at an angle that depends on your latitude. Note that astronomers measure a star's altitude in relation to the horizon (0°), so that if it is at right angles to the horizon (the zenith), its altitude is 90°, and if it is midway between the horizon and the zenith, its altitude is 45°. In the northern hemisphere, Polaris' position corresponds to your latitude and *vice versa*, so that if you are at 60°N, it will be 60° above the horizon. In the southern hemisphere, look for the star Acrux in Crux and work out its altitude. If Crux appears upside down, add 27, and if it is upright, subtract 27; this should give you your latitude: 45° + 27 = 72°S, for example.

PLANISPHERES

A planisphere is a circular map of the stars that is covered with a solid 'mask' – apart from an open section – and is surrounded by dates and times. Different planispheres are sold for different latitudes, so make sure that you buy one that corresponds to yours. Rotate the mask until the correct times of year and night are indicated, and the open section will reveal the stars that are visible in the sky above you. Follow the instructions given for holding the planisphere above your head in order to show the correct horizon, and you'll instantly be able to compare the planisphere's star map with the reality.

THE GLOBAL AND CELESTIAL SPHERES

The Earth is mapped out into an imaginary co-ordinate grid by latitude (horizontal lines) and longitude (vertical lines), and is also divided longitudinally into 24 time zones. Imagine it as a circle: because it turns 360°C in the space of 24 hours, it moves 15° in an hour (360 ÷ 24 = 15). The meridian is centred on Greenwich, England, which has a longitude of 0°, so that Greenwich Mean Time (GMT) equates to mean solar time or Universal Time (UT). When consulting a star map (see below), note that any times given accord to UT; the 24-hour clock is used, with UT beginning at midnight: 00:00 UT (or GMT). If you are in a place that doesn't share Greenwich's meridian, note that it will be one hour later for each 15° west of Greenwich that you are, and one hour earlier by every 15° east that you are. It's advisable to work out your time zone's UT correction

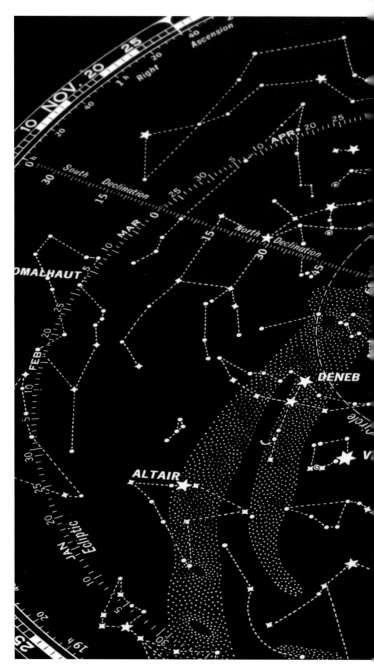

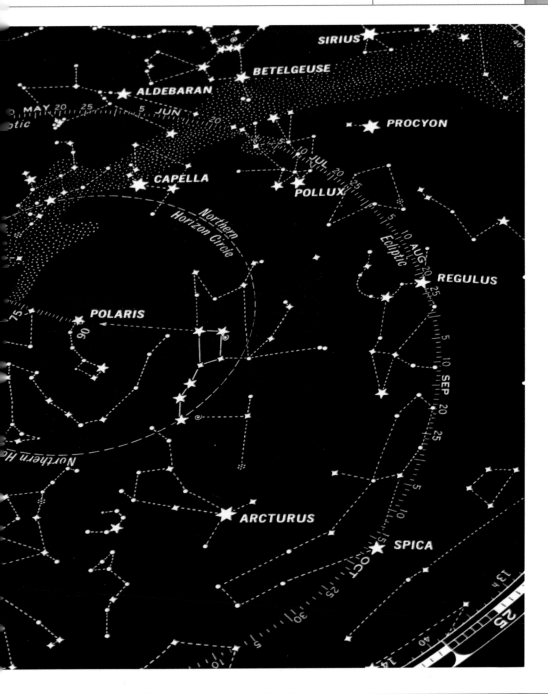

and to practise performing the necessary calculations until they become second nature and you are effortlessly thinking in UT, just like the rest of the worldwide community of astronomers.

Star maps are extremely useful guides to the night sky, but before using them, it's important to understand the rules according to which they are drawn up.

You've probably seen a globe, or a spherical, 3-D representation of a map of the Earth. The sky can be mapped in a similar way, namely as a celestial sphere, or an imaginary sphere surrounding the Earth, with the stars positioned on the inside of the sphere and the Earth at the core. The celestial sphere is marked with an equator (the celestial equator) and two poles (the north celestial pole and the south celestial pole), which are always situated above their counterparts on Earth.

The celestial sphere is also divided up by means of a co-ordinate grid of horizontal and vertical lines. The horizontal lines indicate declination, and the vertical lines signify right ascension. Declination and right ascension are called the celestial co-ordinates. The starting point for measuring declination (which is expressed in degrees, arcminutes and arcseconds) is the celestial equator (0°); north of the celestial equator, positions are given positive (plus) values, while south of it, positions are accorded negative (minus) values. Right ascension (RA) is measured in hours (h), minutes (m) and seconds (s), from 0 to 24 hours; the starting point is the meridian, or the vernal equinox, and the measurements then proceed eastwards.

Star maps are flattened-out, slightly overlapping, versions of the celestial sphere, which is divided up into six sections.

1. One section shows the equatorial constellations that are visible from December to February.

2. A second section illustrates the equatorial constellations that can be seen from March to May.

3. A third section charts the equatorial constellations that can be viewed from June to August.

4. A fourth section shows the equatorial constellations that can be seen from September to November.

5. A fifth section represents the north polar stars, or north circumpolar constellations, or those that are visible all year round in the northern hemisphere: Ursa Major, Ursa Minor, Draco, Cepheus, Cassiopeia and Camelopardalis.

6. A sixth section shows the south polar stars, or south circumpolar constellations, or those that can be seen throughout the year in the southern hemisphere: Crux, Centaurus, Triangulum Australe, Pavo, Tucana, Octans, Hydrus, Dorado, Volans and Carina.

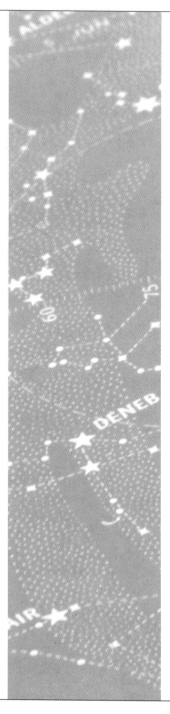

Declination values are printed at the sides of the map, while RA values appear along the top and bottom. The labels 'north horizon', 'west horizon', 'south horizon' and 'east horizon' may be printed on the sides. Remember that these directions refer to the celestial sphere, not to the Earth, so that north is always the direction of the north celestial pole, and if you were looking southwards along the meridian, west would appear on the right, and east on the left.

A good star map should be accompanied by a key. Magnitudes, double stars and variable stars are indicated by different graphic stars and circles. Open clusters, globular clusters, nebulae, planetary nebulae, supernova remnants and galaxies are usually identified by different yellow symbols. The ecliptic is shown as a dotted line. The Milky Way is shown as an amorphous, cloud-like shape. The stars that make up the constellations are linked by lines (rather like a completed dot-to-dot puzzle) which makes them easier to identify. Constellation boundaries may also be shown.

Many newspapers print a star map at the start of a new month, illustrating the stars that will be visible above the horizon on the meridian at 22h (10p.m.) – or 23h (11p.m.) if it is summer time (see page 224 for an explanation of UT) – over the next 30 days (30d) or so. The observer should turn the map so that the horizon that he or she is facing appears at the bottom.

The Eagle Nebula in the constellation Aquila.

Viewing the night sky in the northern hemisphere

Northern-hemisphere star maps show the stars for latitude 45°N, with the observer facing south. Stars in the west will be visible earlier than 22h, and those to the east, later.

Before consulting a star map, you'll need to get your bearings. If it is between November and March and you are in the northern hemisphere, look for Orion, which rises in the east and appears tilted as it moves from left to right across the sky, straightening up at its due-south zenith and then tilting in the opposite direction as it moves towards its setting point on the western horizon. Also look for the distinctive shape of Ursa Major.

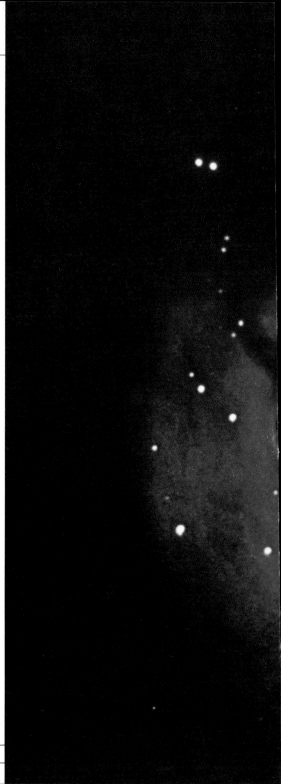

A nebula in Orion.

The Lagoon Nebula in the constellation of Sagittarius.

Remember that the observer is imagined looking south on a northern-hemisphere star map, so that the stars at the top are overhead and the stars at the bottom can be seen on the southern horizon.

Viewing the night sky in the southern hemisphere

Southern-hemisphere star maps show the stars for latitude 45°S, with the observer facing north. Stars in the west will be visible earlier than 22h, and those to the east, later.

If you are in the southern hemisphere, you can also use Orion to help you to get your bearings between November and March. It rises in the east and appears tilted as it moves from right to left across the sky, straightening up at its due-north zenith and then tilting in the opposite direction as it moves towards its setting point on the western horizon. Look out for the cross described by Crux, too.

Remember that the observer is imagined looking north on a southern-hemisphere star map, so that the stars at the top are overhead and the stars at the bottom can be seen on the northern horizon.

SOME SIGHTS TO LOOK OUT FOR DURING THE YEAR

Date	Constellations and other sights
December to February	Pleiades; Auriga; Taurus; Orion; Orion Nebula; Crab Nebula; M35; Gemini; Canis Minor; Canis Major; Sirius; Monoceros; Cancer
March to May	Hydra; Leo; Sextans; Crater; Corvus; Sombrero Galaxy; M61; Virgo Cluster; Coma Berenices; Black Eye Galaxy; Virgo; Boötes
June to August	Milky Way; Cygnus Rift; the Dumbbell Nebula; Scorpius; Serpens Caput; Ophiuchus; Lagoon Nebula; Triffid Nebula; Eagle Nebula; Omega Nebula; Sagittarius; Aquila
September to November	Capricornus; Saturn Nebula; Aquarius; Helix Nebula; Piscis Austrinis; NCG 253; NGC 246; Cetus; M77; M15; Pegasus; Andromeda; Andromeda Galaxy; Triangulum Galaxy; Triangulum

TURN TO PAGE . . .

If you can't identify a bright star that appears close to the ecliptic, it is probably a planet. For further details on viewing Mercury, see pages 68–9; Venus, see page 72; the Moon, see pages 79–84; Mars, see page 89; Jupiter, see page 94; Saturn, see page 101; Uranus, see page 110; and Neptune, see page 114. For advice on viewing asteroids, see page 125; and for comets, see pages 129–30.

Glossary

Familiarise yourself with the terms that astronomers use, and you'll soon be talking the same language!

Absolute magnitude: a star's magnitude at 10 parsecs (32.6 light years) from Earth; a celestial body's real brightness. *See also* magnitude.

Absolute zero: −273.15°C (−459.67°F), or the theoretically lowest-possible temperature.

Absorption line or band: when a spectrum is analysed by a spectroscope, a dark line caused by atoms absorbing radiation. *See also* spectral lines.

Altazimuth: an instrument that measures the altitude and azimuth of a celestial body; an altazimuth mounting enables a telescope to be moved around both a vertical axis (in altitude) and a horizontal axis (in azimuth).

Altitude: the angular distance of an object in the sky measured vertically from the horizon.

Angular measurement: measurement in angles, i.e., in degrees (°), arcminutes (') and arcseconds (").

Antimatter: a form of matter that comprises subatomic antiparticles that contain antiprotons and positrons; matter that has the opposite qualities to conventional matter.

Aperture: an opening, or the diameter of that opening, in an optical instrument that lets a certain amount of light, or radiation, in and out.

Aphelion: in a planet or comet's orbit, the point when it is furthest away from the Sun. *See* perihelion.

Apogee: in the Moon's (or another natural or artificial satellite's) orbit, the point at which it is farthest away from the Earth. *See* perigee.

Arcminute: a unit of angular measurement consisting of 60 arcseconds; the symbol for an arcminute is '.

Arcsecond: a unit of angular measurement equivalent to one 3,600th of a degree (or $\frac{1}{3,600}$

degrees) of an angle, used to calculate the separation or size of objects in the sky; the symbol for an arcsecond is ". *See also* parsec.

Asteroid: a rocky or metallic object in space with a diameter ranging from between a few metres and over 930 km (578 miles). Also called a minor planet or planetoid.

Asteroid Belt: the region that contains 90 per cent of all asteroids, which lies between the orbits of Mars and Jupiter.

Astronomical unit (AU or au): a measurement of distance equivalent to 149.6 million km (92,959,671 miles), or the average distance between the Earth and the Sun.

Atmosphere: a layer of gas surrounding a planet that is kept in place by the planet's gravity; a star's outer layers.

Atom: the smallest quantity of an element that can take part in a chemical reaction, consisting of the subatomic particles protons, neutrons and electrons.

Aurora: when viewed from Earth, a red, green or yellow glow in the sky above the polar regions, often resembling a streamer, band or curtain, caused by a collision between gases in the Earth's atmosphere and charged particles from the Sun that have become trapped within the Earth's magnetic field. The Aurora Borealis (Northern Lights) is seen around the North Pole; the Aurora Australis (the Southern Lights) is seen above the South Pole.

Axis: an imaginary line passing through the centre of a planet or star around which the object rotates.

Azimuth: the angular distance of an object in the sky measured clockwise from the south point of the horizon.

Binary star: a double-star system in which two stars following different orbits revolve around a centre of balance that is common to both. Also known as a binary system.

Black hole: a region in space created by the death and gravitational collapse of a star, triggering a gravitational pull so strong that no matter or radiation (not even light) that has been sucked into it can escape.

Blazar: when viewed from the Earth, an active galaxy that is visibly emitting variable-light radiation from its core.

Blueshift: in a spectrum, a shift of the spectral lines towards the blue end, indicating that the source of the radiation is moving towards the observer. The opposite of redshift.

Brown dwarf: a dark, heat-generating celestial body that is larger than a planet, but smaller than a star.

Celestial latitude: a celestial body's angular distance north or south of the ecliptic.

Celestial longitude: a celestial body's angular distance measured from the vernal equinox eastwards along the ecliptic.

Celestial sphere: an imaginary sphere surrounding the universe on whose surface celestial bodies appear to lie.

Cepheid variable: a class of variable stars whose regular cycles of variation in luminosity and size are related to their absolute magnitudes, and that are therefore used as a basis for measuring distances in space. *See also* variable star.

Chromosphere: a layer of gas in the Sun's lower atmosphere, between the photosphere and corona, that is visible during a total solar eclipse, when it is pink-red in colour.

Circumpolar: circulating around a pole.

Comet: a celestial object made up of ice, snow, rock debris and dust that orbits the Sun, usually within the Oort Cloud; should it enter the inner solar system, the ice and snow in its nucleus will melt and vaporise, causing the comet's head to glow (as the coma) and two luminous tails of dust and gas to stream out behind it.

Conjunction: when viewed from the Earth, the position of a planet or the Moon when it is aligned with the Sun; if an inner planet or the Moon is between the Earth and the Sun, it is termed an inferior conjunction, but if the Sun separates the planet or the Moon and the Earth, it is a superior conjunction; alternatively, the apparent coincidence in position of two celestial bodies on the celestial sphere, i.e., when their celestial longitude is identical.

Constellation: a region of the celestial sphere that contains one of 88 groups of stars visible from the Earth.

Core: the central part of a planet below the mantle; in the Earth's case, it consists primarily of iron and nickel.

Corona: the hot, outer layer of the Sun's atmosphere that is visible as a halo during a solar eclipse; alternatively, a halo of light surrounding any luminous celestial body. Also called the aureole.

Cosmic dust: microscopic particles of matter in the interstellar medium that absorb sunlight. Also called stardust.

Cosmic rays: particles of rapidly moving, high-energy radiation comprising such atomic nuclei as protons that travel to the Earth from outer space. Also called cosmic radiation.

Crater: a bowl-shaped depression on the surface of the Moon or a planet caused by the impact of a meteoroid.

Crust: the solid, rocky outer surface of the Earth, Moon and certain planets.

Culmination: a celestial body's highest or lowest altitude as it crosses the meridian.

Cusp: one of the points of the crescent Moon or of an inferior planet that similarly displays crescent phases.

Dark matter: invisible, undetectable matter that is thought to make up more than 90 per cent of the universe's mass. *See also* WIMP.

Declination: the angular distance of a celestial body, measured in degrees, arcminutes and arcseconds on the celestial sphere, from the celestial equator, either northwards (positive) or southwards (negative), up to +90° and down to –90°; corresponds to latitude on the Earth. *See also* right ascension.

Deep-sky objects: nebulae, star clusters and galaxies.

Degree: the basic unit of angular measurement; there are 360 degrees (°) in a circle; a degree consists of 60 arcminutes.

Direct motion: west-to-east motion.

Direct rotation: the anti-clockwise rotation of a planet or satellite when above the North Pole.

Doppler effect: the effect by which the apparent frequency of waves of electromagnetic radiation (or sound) changes when they reach the observer if the source is either moving nearer (blueshift) or further away (redshift). Also called the Doppler shift.

Eclipse: the obscuring of one celestial body by another passing in front of it when viewed from the Earth; an eclipse may be partial, total, or, in the case of the Sun, annular (when a ring of sunlight remains visible around the lunar disc); a solar eclipse occurs when the Moon passes between the Sun and the Earth; a lunar eclipse occurs when the Earth passes between the Sun and the Moon.

Ecliptic: on the celestial sphere, the annual path that the Sun appears to follow across the sky in relation to the constellations; it is inclined at an angle of 23.45° to the celestial equator.

Electromagnetic radiation: radiation, or waves of energy, that comprises electric and magnetic fields at right angles to one another; it travels through space at the speed of light, carried by photons. The electromagnetic spectrum ranges from long radio waves and microwaves through infrared radiation, visible light, ultraviolet radiation and X-rays to short gamma rays.

Electron: a stable, negatively charged, elementary particle orbiting the nucleus of an atom.

Elongation: when viewed from the Earth, the point at which an inferior planet is the furthest, in terms of angular distance, from the Sun.

Emission line: on a spectrum, a bright line whose source is energy emitted by atoms at a specific wavelength. *See also* spectral lines.

Equinox: on the vernal and autumnal equinoxes, which occur six months apart when the Sun crosses the plane of the equator, the day and night are of equal length; alternatively, one of the two points (equinoctial points) on the celestial sphere where the ecliptic and celestial equator cross. The vernal equinox (also called the first point of Aries) now occurs in the constellation of Pisces, and the autumnal equinox (or the first point of Libra), today occurs in the constellation of Virgo. *See also* precession.

Extrasolar: outside the solar system.

Eyepiece: in an optical instrument, the magnifying lens to which the viewer applies his or her eye in order to view the image produced by the main optic, or objective.

Filament: a string of galaxy superclusters; giant solar prominences.

Focal length: in an optical instrument, the distance between the centre of a lens or reflecting surface of a mirror and the point at which the light rays converge and the image is brought into focus (i.e., the focal point).

Following: a position to the east of an object. *See also* preceding.

Frequency: in terms of electromagnetic radiation, the number of waves that pass a certain point every second.

Galaxy: a deep-sky object comprising star systems (and sometimes nebulae, dust and gas) held together by gravitational attraction. Depending on their structure, contents and behaviour, galaxies can be classified as elliptical, lenticular, spiral, barred spiral and

irregular; they may also be described as being starburst, active, radio or Seyfert galaxies. *See also* blazar and quasar.

Galaxy cluster: a group of galaxies kept together by gravity. *See also* supercluster.

Gamma rays: rays of gamma radiation, a form of electromagnetic radiation with very short wavelengths.

Gas giant: a planet that consists primarily of a dense gaseous atmosphere; Jupiter, Saturn, Uranus and Neptune are gas giants.

Giant star: *see* red giant.

Gibbous: when more than half of the sunlit face of the Moon, for example, is visible from Earth, but not the entire face.

Gravitational lensing: the multiple images seen when viewing a distant object, such as a galaxy, due to the distortion of light as it passes through an area of powerful gravity.

Gravity: the force of attraction that bodies with mass exert on one another. Also known as gravitation.

Halo: a ring of light around a celestial body, for example, the Sun and the Moon; alternatively, a spherical cloud comprising globular star clusters or dark matter that surrounds spiral galaxies.

Heliosphere: the Sun's boundary, or the most distant area that solar wind influences, calculated to be within 100 AU of the Sun.

Helium (He): the second most common element in the universe after hydrogen, a light, inert gas.

Hertzsprung–Russell (HR) diagram: a graph that plots the spectral type (temperature and hence colour) of stars against their visual luminosity (absolute magnitude, or brightness), which can be used to track the evolution of stars.

HST: Hubble Space Telescope.

Hubble constant: the rate at which the universe is expanding, which is currently believed to be 73 km (around 45 miles) per second per megaparsec distance.

Hydrogen (H): the lightest and most plentiful element in the universe, a flammable, colourless gas.

IAU: the International Astronomical Union, founded in 1919, one of whose functions is to assign names to newly discovered celestial bodies.

IC number: 'IC' stands for 'Index Catalogues', which are catalogues listing celestial bodies by number; a celestial object may consequently be referred to by its IC number.

Inferior planet: Mercury and Venus, whose orbits fall within that of the Earth and are therefore 'inferior' to it.

Infrared rays: on the electromagnetic spectrum, rays whose wavelengths or frequencies are longer than those of visible light, but shorter than those of microwaves and radio waves.

Inner planets: Mercury, Venus, the Earth and Mars, the planets nearest the Sun, which are separated from the outer planets by the Asteroid Belt.

Intergalactic: between galaxies.

Interstellar: between stars.

Interstellar medium (ISM): the matter that lies between the stars, which is mainly made up of clouds of ionised, neutral or molecular hydrogen.

Ionosphere: an electron-rich section of the

Earth's atmosphere about 60–1,000 km (37–621 miles) above its surface.

Kiloparsec: 1,000 parsecs.

Kuiper Belt: in the solar system, the region between Neptune's orbit and the Oort Cloud that contains frozen objects – Kuiper Belt objects, or KBOs – that resemble comets. Also known as the Edgeworth–Kuiper Belt.

Light: anything that illuminates; on the electromagnetic spectrum, radiation whose wavelengths are visible to the human eye and whose frequency falls between that of infrared and ultraviolet.

Light, speed of: the distance that a ray of light travels in a second, namely 300,000 km (186,417 miles), and a standard unit of astronomical measurement; nothing can move faster than this.

Light year (ly): the distance that a ray of light theoretically travels in a year, namely 9,500 trillion km (5,903 trillion miles), and a standard unit of astronomical measurement.

Limb: when viewed from the Earth, the edge of the Sun, the Moon or another planet's apparent disc.

Lithium (Li): a silvery element and the lightest (alkali) metal.

Local Group: the cluster of galaxies that includes the Milky Way.

Luminosity: the amount of light, or energy, that a star radiates per second.

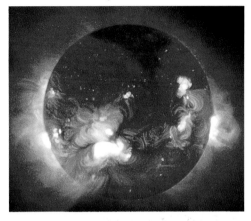

Lunation: the amount of time that it takes the Moon to travel around the Earth between two successive new Moons, which is about 29 days, 12 hours, 44 minutes and 3 seconds; the cycle of lunar phases. Also known as a lunar or synodic month.

Magnetosphere: the area around a planet containing charged particles that are influenced by the planet's magnetic field and can consequently resist the solar wind.

Magnitude: the apparent brightness of a celestial body as viewed from the Earth and expressed on a numerical scale, with the brightest having a low (and even negative) number, and the dimmest having a high, positive number; each integral value represents a brightness about 2.5 times greater than the next highest. Also called apparent magnitude. *See also* absolute magnitude.

Main Belt: *see* Asteroid Belt.

Main sequence: with reference to a Hertzsprung–Russell diagram, the diagonal band that contains roughly 90 per cent of all known stars; main-sequence stars generate energy, so that as it evolves during its lifetime, a young star will move onto the main sequence, spend 90 per cent of its lifespan on it and eventually drop off as it fades and dies.

Mantle: the layer of rock that separates a planet's core and crust.

Mare: the Latin word for 'sea' (plural 'maria') that is the basis for the names of a number of enormous, lava-filled depressions on the Moon (and Mars) that resemble dark seas when viewed from the Earth.

Mass: a measure of the amount of matter in a body and consequently an indication of how strongly it can resist the forces of gravity (gravitational mass) and velocity (inertial mass).

Matter: any substance that has mass and occupies space, i.e., has a physical presence.

Megaparsec: 1 million parsecs.

Meridian: with reference to the celestial sphere, the great circle that runs through the north and south celestial poles and the observer's zenith and nadir; the great circle that runs through a celestial body's north and south poles, as well as the observer's position.

Meteor: the bright streak of light generated by a tiny meteoroid that has entered the Earth's atmosphere and is burning up. This phenomenon gives rise to the names 'shooting star' and 'falling star'.

Meteor shower: a regularly occurring 'shower' of meteors that occurs when the Earth passes through a 'swarm' of meteors.

Meteorite: the remains of a meteoroid that has landed on a planet or moon's surface. The force of the impact may sometimes create a crater.

Meteoroid: a small fragment of an asteroid or comet that consists mainly of rock and dust and is thought to orbit the Sun.

Methane (CH⁴): a colourless, flammable gas that is made up of carbon and hydrogen.

Microgravity: the effects of the operation of gravity on a localised area.

Microwaves: on the electromagnetic spectrum, the wavelengths or frequencies of microwaves fall between radio waves (of which they are a type) and infrared radiation; microwaves are used in radar and as carrier waves in radio broadcasting.

Minor planet: *see* asteroid.

M number: one of 103 numbers assigned by Charles Messier to the same number of fuzzy-looking objects in the sky; for instance, in the Messier Catalogue, the Andromeda Galaxy's M number is 31, so that its alternative name is M31.

Molecular cloud: a cool, dense interstellar region consisting of a cloud of various molecules – but mainly hydrogen molecules – and dust, in which stars are being born.

Molecule: two or more atoms held together by chemical bonds that behave as one; the simplest chemical compound.

Moon: a planet's natural satellite. (In this book, 'the Moon' describes the Earth's satellite, while 'moon' or 'moons' refer to other planets' natural satellites.) *See also* satellite.

Nadir: on the celestial sphere, the point directly below the observer and opposite the zenith.

Nebula: (plural, nebulae or nebulas) a cloud of gas (primarily hydrogen) and particles (mainly of dust) that manifests itself as a fuzzy patch of light to an observer on the Earth. Categories of nebulae include emission, reflection and dark nebulae. *See also* planetary nebula.

Neutrino: a neutral elementary particle produced by nuclear fusion that has next to no mass and travels at the speed of light.

Neutron: a neutral elementary particle in the nucleus of an atom.

Neutron star: a star that has collapsed under its own gravity, often after a supernova, that consists almost exclusively of neutrons.

NGC number: 'NGC' stands for New General Catalogue [of Nebulae and Clusters of Stars], which was first published in 1888; a nebula, star cluster or galaxy may be known by its NGC number, which is the number allocated to it in this catalogue of more than 8,000 deep-sky objects.

Nitrogen (N): a colourless, gaseous element that makes up 78 per cent of the air in the Earth's atmosphere.

Nova: (plural, novae or novas) a white dwarf in a binary system will attract material from its fellow star; this material builds up into an atmosphere, which eventually ignites, creating a nova that burns fiercely and brightly before gradually fading. *See also* supernova.

Nuclear fusion: when two atomic nuclei combine to form a heavier nucleus in extremely hot and pressurised conditions, releasing energy in the process. This nuclear, or atomic, energy creates and fuels stars.

Nucleus: (plural, nuclei) the positively charged centre of an atom, comprising protons and neutrons and orbited by electrons; the centre of a comet's 'head'.

Objective: the main image-forming lens or mirror in an optical instrument.

Occultation: the temporary disappearance from view of a celestial body when another passes in front of it.

Oort Cloud: a 1.6-ly-wide, spherical region beyond Pluto that is densely packed with billions of comets orbiting the Sun. It is named for its proposer, the Dutch astronomer Jan Oort (1902–92).

Opposition: on the celestial sphere, the position of two celestial bodies that are diametrically opposite one another; when viewed from the Earth, the position of an outer planet when it is directly opposite the Sun, when it is best observed; when viewed (hypothetically) from the Sun, the position of an outer planet or the Moon when it is in line with the Earth.

Orbit: the elliptical path (an elongated circle) followed by a celestial body as it moves around another, more massive, celestial body under the influence of its gravitational pull.

Orbital period: the time that it takes a celestial body to orbit another.

Outer planets: Jupiter, Saturn, Uranus, Neptune and Pluto, all of which lie beyond the Asteroid Belt.

Oxygen (O): a colourless, reactive, gaseous element that makes up 20 per cent of the Earth's atmosphere.

Parallax: the apparent shift of an object's position against a distant background when the observer moves his or her own position, i.e., when it is viewed from two different points; the angle through which a star moves during six months. Working from opposite points on the Earth's orbit, astronomers use parallax to calculate the distances of stars.

Parsec: or *par*allax *sec*ond in full, a unit of distance equal to the distance from the Earth at which a celestial body has a parallax angle of 1 arcsecond against the sky when the Earth moves one AU around the Sun, equivalent to about 3.26 light years. *See also* arcsecond, kiloparsec and megaparsec.

Penumbra: the outer, lighter area of a sunspot; an area of half-shadow cast by one celestial body onto another.

Perigee: the point in its orbit when the Moon (or an artificial satellite) is nearest to the Earth. *See also* apogee.

Perihelion: the point in its orbit when a planet or comet is closest to the Sun. *See also* aphelion.

Phase: when viewed from the Earth, a shape, such as a crescent, that the Sun-illuminated Moon or an inferior planet regularly displays. The Moon's four principal phases are new Moon, first quarter, full Moon and last quarter. *See also* cusp.

Photon: a particle of electromagnetic radiation with zero rest mass and charge.

Photosphere: the surface of the Sun or another star that is visible from the Earth; a star's point of transparency that enables its light to be seen by Earthlings.

Planet: a spherical celestial body made up of rocky or gaseous material that is in orbit around a star and is illuminated by its light, rather than generating its own light. In our solar system, nine planets revolve around the Sun: Mercury, Venus, the Earth, Mars, Jupiter, Saturn, Neptune, Uranus and Pluto. *See also* inner planets, outer planets, inferior

planet and superior planets.

Planetary nebula: an expanding, light-emitting shell of gas formed by the matter that a red giant discards before it becomes a white dwarf. *See also* nebula.

Planetesimal: a large rocky object. *See also* protoplanet.

Planetoid: *see* asteroid.

Positron: an electron's antiparticle or antimatter, in that it has the same mass as an electron, and is equally charged, too, the difference being that its charge is positive rather than negative.

Preceding: a position to the west of a specific object. *See also* following.

Precession: a wobble in the motion of a spinning body, such as a planet, causing its rotational axis to sweep out to describe a cone. In the case of the Earth, this shift in the point where the Equator and ecliptic cross is caused by the combined effect of the Sun, the Moon and other planets' gravity. It is the precession of the Earth's axis that results in the precession of the equinoxes, that is, the slow, westward shift of the vernal and autumnal equinoxes and their slightly earlier occurrence each year. *See also* equinox.

Prominence: only visible from the Earth during a total eclipse, an incandescent stream of gas that flares out from the Sun's lower corona.

Proper motion: a small, yet sustained, change in a star's direction of motion in relation to the Sun; a star's motion in relation to the celestial co-ordinates.

Proton: a stable, positively charged, elementary particle in an atom's nucleus.

Protoplanet: an embryonic planet that is larger than a planetesimal.

Protostar: an embryonic star, i.e., a cloud of interstellar medium that is slowly collapsing to create a dense, hot core in which nuclear fusion, and then star birth, will eventually occur.

Pulsar: (or, in full, *puls*ating st*ar*) a small, dense, rapidly rotating neutron star that emits regular pulses of polarised radiation, mainly in the form of radio waves.

Quasar: (or *quas*i-stell*ar* object) a very distant, active galaxy, that radiates huge amounts of energy – including visible light – from a small, dense centre.

Radar: (or *r*adio *d*etecting *a*nd *r*anging) a method of detecting the position and velocity of a very distant object by transmitting high-

frequency radio pulses in its direction and then, using a radarscope, noting the direction from which they are bounced back, as well as the time that it takes to receive a reflected radio pulse.

Radiation: *see* electromagnetic radiation.

Radio wave: a wave of electromagnetic radiation at radio frequency, which occurs at

the end of the electromagnetic spectrum where the waves are the longest.

Red dwarf: a low-mass, small, cool, red main-sequence star.

Red giant: a Sun-like star that is in its final evolutionary stage, having increased in size, become brighter and changed colour. Red giants are of the spectral type M and radiate red light. Also known as a giant star. *See also* supergiant.

Redshift: in a spectrum, a shift of the spectral lines towards the red end, indicating that the source of the radiation is moving away from the observer. The opposite of blueshift.

Retrograde motion: movement in the opposite direction to that in which the Earth orbits the Sun; when a superior planet appears to move backwards, i.e., from east to west, when viewed from the Earth due to the Earth 'overtaking' it as both planets orbit the Sun; movement in the opposite direction to any planet's rotational direction.

Right ascension: the angular distance of a celestial body measured eastwards along the celestial sphere's equator in hours (h), minutes (m) and seconds (s), from 0 to 24 hours, from the vernal equinox to the point at which the celestial equator crosses a circle that travels through the celestial body and the celestial pole; corresponds to longitude on the Earth. *See also* declination.

Satellite: a celestial body (a natural satellite) that, due to gravitational attraction, orbits around another; a manmade structure (an artificial satellite) that orbits around a planet, star or moon. *See also* moon.

Shepherd moon: a natural satellite that limits the extent of a planet's ring by gravitational force.

Solar flare: a short, powerful, explosion of high-energy radiation above the Sun's surface, caused when two loops of the Sun's magnetic field collide.

Solar system: a system with the Sun (or another Sun-like star) at its centre, with planets (in our solar system, Mercury, Venus, the Earth, Mars, Jupiter, Saturn, Uranus, Neptune and Pluto), asteroids and comets orbiting it, all held within the Sun's gravitational field.

Solar wind: a stream of high-velocity, charged particles emitted by the Sun.

Spectral analysis: the scientific analysis of a spectrum's spectral lines in order to learn more about the nature of their source.

Spectral lines: on a spectrum, dark or bright lines that indicate that radiation is being absorbed or emitted by their source. *See also* absorption line and emission line.

Spectral type: a star's classification according to its colour and surface temperature and hence spectral lines. The Harvard classification system classifies stars by spectral type into the classes O, B, A, F, G, K and M, for instance. Also known as spectral class.

Spectrometer, spectrograph, spectroscope: instruments used to produce and record a spectrum for analysis.

Spectrum: (plural spectra) the entire range of electromagnetic radiation; a band of radiation from a particular source, divided into different wavelengths and displaying spectral lines; the distribution of colours that becomes visible when white light is split into its components, as seen in a rainbow.

Standard candle: the basic unit of luminous intensity; a celestial object, such as a Cepheid variable, whose brightness is known and that therefore provides a reliable basis for comparisons and calculations. Also known as candela.

Star: a huge, hot, luminous and gaseous celestial object that generates energy by means of nuclear fusion.

Star cluster: an aggregation of stars held together by gravity; star-cluster types include globular clusters, or thousands of old stars crowded together in a globular formation, and open clusters, or less dense groupings of hundreds of young stars.

Subatomic particle: a particle that is smaller than an atom, such as a proton, a neutron and an electron.

Sunspot: a cool, dark area with a powerful magnetic field on the Sun's surface that is relatively short-lived. *See also* umbra.

Supercluster: a group of galaxy clusters

kept together by gravity.

Supergiant: a star whose mass was 10 times that of the Sun before it evolved into a huger version of a red giant with the most intense luminosity of any star in the universe.

Superior planet: Mars, Jupiter, Saturn, Uranus, Neptune and Pluto, all of whose orbits fall outside that of the Earth, making them 'superior' to it.

Supernova: (plural, supernovae or supernovas) when a supergiant runs out of fuel and explodes (a type II supernova), or a white dwarf does (a type Ia supernova), this massive, incredibly bright explosion is called a supernova. *See also* nova.

Terminator: the dividing line between the illuminated and dark sections of the Moon or a planet.

Transit: when viewed from the Earth, the passage of a celestial body or satellite across the face of a larger-seeming celestial body; the apparent passage of a celestial body across the meridian.

Ultraviolet rays: on the electromagnetic spectrum, rays whose wavelengths or frequencies are longer than those of X-rays, but shorter than those of visible light.

Umbra: an area of total shadow cast by one celestial object on another; the dark, inner part of a sunspot.

Variable star: a star whose brightness (and sometimes size) varies irregularly or regularly. *See also* Cepheid variable.

Void: a massive area of empty space between filaments.

Wavelength: the distance between two peaks, or two troughs, in a series of waves of electromagnetic radiation.

White dwarf: a dying, Sun-like star that has stopped producing energy and whose core has collapsed, leaving a faint, small, dense body.

WIMP: (a *w*eakly *i*nteracting *m*assive *p*article) a type of particle created during the big bang; WIMPs are believed to be a major component of dark matter.

X-rays: on the electromagnetic spectrum, rays whose wavelengths or frequencies are longer than those of gamma rays, but shorter than those of ultraviolet radiation.

Zenith: on the celestial sphere, the point directly above the observer. *See also* nadir.

Zero gravity: weightlessness.

Zodiac: an imaginary area extending 8° on either side of the ecliptic that contains the 12 zodiacal constellations, each 30° in extent, through which the Sun, the Moon and the planets appear to move.

Index

X

Z

Clare Gibson is a writer and editor whose life-long interest in astronomy has not been dampened by living in light-polluted London, England. She is the author of a number of books, including two on astrology, the ancient star-gazing tradition that, unusually for an astronomy writer, she does not believe to be fundamentally incompatible with astronomy.

CREDITS

The author dedicates this book to her godson, Jacob Millidge, a star in the making! She also extends her gratitude to Mike Haworth-Maden, as well as to John and Marianne Gibson, for their generosity in supplying advice, information and encouragement.

Bibliography

Cooper, Chris; Spence, Pam; and Stott, Carole, *New Illustrated Stars & Planets*, Selectabook Ltd, Devizes, 2002.

Cornelius, Geoffrey, and Devereux, Paul, *The Secret Language of the Stars and Planets*, Pavilion Books Ltd, London, 1996.

Couper, Heather, and Henbest, Nigel, *Space Encyclopedia*, Dorling Kindersley Ltd, London, 1999.

Dorschner, Johann; Friedemann, Christian; Marx, Siegfried; and Pfau, Werner, *Astronomy: A Popular History*, Almark Publishing Co Ltd, New Malden, 1975.

Dunlop, Storm, *Practical Astronomy*, Philip's, London, 2003.

Ekrutt, Joachim, *GU Kompass Sterne*, Gräfe und Unzer Verlag GmbH, München, 2004.

Goodwin, Simon, *Hubble's Universe*, Constable and Co Ltd, London, 1996.
Gribbin, John, *Space: Our Final Frontier*, BBC Worldwide Limited, London, 2001.

Hoskin, Michael (Ed), *The Cambridge Concise History of Astronomy*, Cambridge University Press, Cambridge, 1999.

Nicolson, Iain, *Astronomy: Exploring the Night Sky*, Treasure Press, London, 1970.

Tirion, Wil, and Dunlop, Storm, *Der Kosmos Sternführer*, Franckh-Kosmos Verlags-GmbH & Co, Stuttgart, 2004.

Useful websites

NASA: www.nasa.gov/home/index.html

The Society of Popular Astronomy:
www.popastro.com

The International Astronomical Union:
www.iau.org

The US Geological Survey:
www.astrogeology.usgs.gov

The European Southern Observatory:
www.eso.org

The Space Telescope Science Institute:
www.stsci.edu/resources

The BBC: www.bbc.co.uk/science/space

Astronomy Now magazine:
www.skyandtelescope.com
www.skypub.com

Views of the solar system:
www.solarviews.com/ss.html

Satellite predictions:
www.heavens-above.com

Picture credits

© Getty Images: pp6, 8, 10-12, 14, 16, 20t, 20br, 21, 22t, 24, 28, 31b, 32, 34, 37-38, 44, 48, 55, 57, 58-59, 60b, 61, 63, 64-67, 70-71, 74, 77, 78-80, 85, 86b, 91, 92b, 93t, 97, 98t, 99b, 100t, 103, 111, 118-119, 124t, 127-129, 134, 136, 137t, 141, 145, 149, 153, 173, 183t, 192, 195, 202-206, 209, 217-218, 220, 222-223, 230, 232.

© Stockbyte: pp20bl, 33, 75, 83, 208, 211-212, 216.

Courtesy NASA/JPL-Caltech: pp19, 22b, 23t, 26-27, 35-36, 39, 42, 46, 50-51, 56, 60t, 62, 68-69, 73, 82, 84, 86t, 87-90, 92t, 93-96, 98b, 99t, 100-101, 104-110, 112-117, 120-121, 124b, 125-126, 131, 137b, 138, 140, 142, 146-148, 151-152, 155-156, 158, 172, 174, 177, 181-182, 183b, 184, 186, 188, 190, 197.

© Corbis: pp13, 15, 18, 23b, 30, 31t, 37t.

Image p178 © Dennis Glory

Images pp224-227 © Jan Tyler

(Where t = top, b = bottom, l = left and r = right)